T. Longmore

A Treatise on Gunshot Wounds

T. Longmore

A Treatise on Gunshot Wounds

ISBN/EAN: 9783337021542

Printed in Europe, USA, Canada, Australia, Japan

Cover: Foto ©berggeist007 / pixelio.de

More available books at **www.hansebooks.com**

LONGMORE

ON

GUNSHOT WOUNDS.

A TREATISE

PHILADELPHIA:

J. B. LIPPINCOTT & CO.

1862.

CONTENTS.

GUNSHOT WOUNDS IN GENERAL.

(v)

GUNSHOT WOUNDS.

GUNSHOT WOUNDS consist of injuries from missiles projected by the force of explosion. As the name implies, this class of wounds is ordinarily restricted to injuries resulting from fire-arms; but it should be remembered that wounds possessing the same leading characteristics may result from objects impelled by any sudden expansive force of sufficient violence. Injuries from stones, in the process of blasting rocks, or from fragments of close vessels burst asunder by the elastic power of steam, offer familiar examples of wounds of a like nature with those from gunshot. In the following article, however, gunshot wounds will be considered as they are met with in the operations of warfare.

HISTORY.

From the earliest time of the application of gunpowder to implements of war, down to the present day, the wounds inflicted by its means have excited the most marked interest among surgeons; nor can this be wondered at, when the immensely superior energy of this agent in comparison with all the mechanical powers previously in use for hostile purposes, and the terrible nature of its effects on the human frame, are remembered. By its introduction the whole aspect of war was changed, in a great degree, by the distance at which opposing forces were enabled to contend with

(9)

each other; just as, in our day, the nature of battle seems
destined to undergo another change from the increased
range and precision of fire obtained through the general use
of rifled weapons. But though the alterations now being
made in the qualities of fire-arms are of the utmost importance to those whose business and especial study is the art of
war, to the army surgeon the interest they excite is chiefly
limited to the degree of injury and destruction inflicted by
them as compared with weapons of a less perfect kind;
while to the surgeons employed at the time of the introduction of gunpowder, the wounds were wholly new in their
nature as well as degree. Recollecting the ignorance which
then prevailed in all departments of science and art, it can
excite no surprise that the new engines of war, with the
flame and noise accompanying their discharge, were regarded with superstitious terror; nor that surgeons for a
long time found an explanation of the sloughing severity of
the injuries they inflicted, and of their difficult cure, in the
poisonous nature of gunpowder, or of the projectiles which
had been acted upon by it, or in the burning effects of these
latter from heat acquired in their rapid flight through the
air. Unfortunately, these erroneous views did not end with
the theories from which they started, but led to treatment
which only aggravated the evils inflicted by the new weapons,
and interrupted the progress of the healing action, which
nature would otherwise have established. The wound being
regarded as a poisoned wound, it was only by a long and
tedious process of suppuration that the poison could be
hoped to be got rid of from the surface, and prevented from
entering the system of the patient. The irritative fever, the
wasting and emaciation, and all the other results of the protracted cure of the injury were so many evidences of the
indirect effect of the poison working in the frame; just as
the constitutional shock at the time of the wound, the loss
of vitality along the surface in the track of a small projectile, or of the tissues laid bare by the passage of the cannon-

ball were regarded as evidences of its direct influence. On looking back at the works of successive writers on this class of injuries, the reader is surprised that the improvement in their treatment has been so gradual and slow; and cannot fail to observe that the chief impediment to a more rapid amelioration of the system pursued has been the prevailing idea of the necessity of delaying the tendency of nature to close the wound, in order that the supposed poison might be eliminated from the constitution. The openings of entrance and exit and track of the ball were incised; the wound dilated by tents or other means, and terebinthinates, or even boiling oil, poured into it; irritating compounds and ointments applied where superficial dressings were practicable; and it was only after the wound was considered to be fully purged of its venom and foul humors by the extensive suppurative action thus kept up, that cicatrization was permitted to be established.

It required long years of observation in many conflicts, and the exercise of much industry, not to mention moral courage in opposing authorized custom and prejudice, before a simpler and more rational mode of practice was followed. It is satisfactory to know that though Continental surgeons have written more voluminously on the subject of gunshot wounds, the older English military surgeons and writers stand forth conspicuously in leading the way to a more practical knowledge of their nature and proper treatment.

Although, however, much that was erroneous was removed by the earlier surgeons, the light of science can hardly be said to have penetrated this important province of military surgery until the great and last work of John Hunter, on the Blood, Inflammation, and Gunshot Wounds, was published in 1794. This distinguished philosopher filled some of the highest positions in the British service, having been appointed in 1776 Surgeon Extraordinary to the Army, in 1786 Deputy Surgeon-General, and subsequently Surgeon-General; but he only served abroad about three years, and then only had the opportunity of seeing active service as staff-surgeon in the expedition to Belleisle. Had the field of his practical observation been more extensive, there can be no doubt that his zealous and scientific mind would have turned the advantage to the most valuable results for humanity. The physiological principles which he enunciated, based on extensive study and observation in civil life, cannot be controverted; but their practical application, so far as regards the treatment of gunshot wounds, has been greatly modified since his treatise on the subject was published. There cannot be a better illustration of the special position in which this department of military surgery is placed, from the peculiar circumstances under which it is practiced, than the fact that, though men of the highest mental attainments have discussed the subject of gunshot wounds, we are nevertheless indebted to practical experience in military campaigns for every improvement, some few of recent date excepted, that has occurred in their treatment. Thus John Hunter was led to advocate very strongly the delay of amputation, after severe gunshot wounds, for weeks, that the patient's constitution might accommodate itself to the injury; while more extended observation has demonstrated that such secondary amputations are more fatal than those which are performed shortly after the infliction of the wounds leading to them—the advantage of the patient thus coinciding with what must very constantly happen to be a practice of necessity in the field.   Mr. Guthrie remarks, in his Commentaries on the Surgery of the Peninsular War, between 1808 and 1815, that the surgical principles and the practice which prevailed at the commencement of the war were superseded on almost all important points at its conclusion; and he quotes a remark of Sir Astley Cooper to the effect that the art of surgery received from the practical experience of that war an impulse and improvement unknown to it before.

The still more recent military operations in Algeria, in Sleswick-Holstein, in the Crimea, and in India have afforded the opportunity of testing practically the applicability to army practice of some of the great improvements which have been accomplished in the civil practice of surgery in Europe since the termination of the war in 1815. Among these may be particularly enumerated the avoidance of amputation of limbs by recourse to excision of joints; resections of injured portions of the shafts of long bones; mitigated amputations, by removal only of those terminal portions of the extremities which had been destroyed by the original injury; and the practice generally of what has been styled conservative surgery. In these wars, too, the value of chloroform as an anesthetic agent in military surgery has been fully established. They have also especially illustrated the influence of various states of health and climates on the results of gunshot wounds. All the anticipations which were held out at the commencement of some of these campaigns have not been realized, but still they have added much valuable information and many improvements to military surgery.

The alterations made during the last five or six years in the arms of a great proportion of the troops of the leading powers of Europe, and which will, no doubt, be extended to all soldiers in regular armies—namely, the transformation of muskets into "*armes de précision*," with rifled barrels and graduated aims—have led to changes in the severity and almost in the nature of gunshot wounds from small balls; and the consideration of these changes requires the especial attention of army surgeons. The effects of the new rifle-balls were widely witnessed during a portion of the period of the Crimean war. The campaign just concluded in Italy will probably produce additional practical observations from the Continental surgeons engaged in it. The fearful proportion of killed and wounded—greater than in any former experience—will have shown the effects not only of rifled muskets, but of rifled cannon also; and in the French forces engaged an opportunity will have been afforded of institu-

2*

ting a comparison of the results of their treatment under circumstances of bodily health and hospital accommodation very different from those of the French army in the Crimea. It may be hoped that the experience thus gained will advance the knowledge of gunshot wounds and their treatment a still further stride toward accuracy.

In England, one valuable result which emanated from the late war with Russia was the regular collection and arrangement, under government authority for the first time, of the observations and practice of the medical officers employed in the campaign. The value to science of such systematized historical records, if fairly and fully developed, can scarcely be overrated ; and it is to be hoped that henceforth a similar course will be always adopted whenever the country may become involved in war.

VARIETIES OF GUNSHOT WOUNDS.

Gunshot wounds are modified in their nature by the form and kind of missile, by the degree of force with which it is propelled, and by the seat of injury. They are, in addition, affected by the circumstances in which the soldier happens to be placed, and by the state of his health when the injury is received.

Form and nature of missile. — The projectiles used in warfare of the present day are cannon and musket shot, shells of various kinds, hand grenades of iron or thick glass, case-shot, slugs, and other minor varieties of such missiles. These are the ordinary instruments of *direct* gunshot wounds in warfare ; but, in addition, there are numerous sources of *indirect* wounds, resulting from the discharge of cannon and musketry. These are stones, or other hard substances, struck from parapets or from the surface of the ground by cannon-shot ; splinters of wood from platforms and framework, or of iron from gun-carriages ; fragments of bone from wounded comrades, or articles in their possession ; and any other miscellaneous objects which may happen to come into contact with the solid ball or shell in its course. The objects above enumerated present several varieties of forms. The chief are—1st, spherical, as cannon-balls, grape, musket-shot, and shells ; 2d, cylindro-conoidal, as balls belonging to rifled cannon and rifled muskets ; 3d, irregular, but generally bounded by linear and jagged edges, as frag- ments of shells and splinters.

A gunshot wound, whether received from a direct or indirect projectile, may be complicated by the entrance of extraneous bodies of various kinds, most commonly portions of the cloth or buttons of the dress worn by the person wounded. Such foreign substances, though not of themselves causing the wound, often have a special bearing on the progress of its cure.

Not only the form of outline, but the weight, and in some instances the matter of which the missile is composed, influence the nature of gunshot wounds. In the largest kinds of balls, such as are projected from field-pieces or guns of position, the form offers little subject for consideration to the surgeon. So long as there is momentum enough to carry forward the mass of iron of which these missiles are composed, so long will their weight be the most important ingredient in the production of the wounds inflicted by them. Whether the shot come as a solid cone or bolt from one of the new guns or as a round ball from an ordinary cannon, the injury will be equally destructive to life or limb. The same remark is applicable to the heavier forms of shell, before explosion. The only difference surgeons may look for from the use of cylindro-conoidal balls, or Whitworth bolts applied to cannon, should they become general, independent of increase in the number of direct wounds from greater power and precision of fire, will be the less number of indirect injuries likely to result from their action, as they neither ricochet nor roll as "spent balls" in the manner that spherical shot are accustomed to do.

Grape-shot, canister, and spherical case, on striking col-
lectively—that is, before they have spread—as sometimes
happens in assaulting or in accidental close proximity to
guns in the field, produce the same kinds of injuries as cannon-shot, but individually resemble musket-shot in their
effects. Wounds from grape-shot are always of a grave
character, not only from the extent of the flesh wound, but
also because, from their large diameter and weight, the
nerves and vessels of the part struck are less likely to escape
injury, if not destruction, than in wounds from the smaller
shot projected in canister or spherical case.

With regard to musket-shot, the form presents several
features for the consideration of the military surgeon. In
discussing the subject, however, it must not be omitted to be
borne in mind that we have no experience of the effects of
round musket-balls propelled with the same amount of force
as recent improvements in fire-arms have given to balls fur-
nished with a conical vertex; although, in the old, two-
grooved rifle, with its belted round ball, a momentum was
procured far exceeding that of the common smooth-bore
musket. The change in form from the round to the prolonged cylindro-conoidal ball seems to derive its chief importance in surgery from the conical end possessing the
mechanical characteristic of a wedge, while the former acted
simply as an obtuse body. From this quality the power of
penetration of conical bullets is greater, independent of the
increased momentum communicated to them by the construction of the weapons from which they are discharged. Thus,
supposing one of the old musket-bullets to strike a limb at
80 yards, and an Enfield rifle conical bullet of the same
weight at 800 yards, the rate of velocity being similar in
each case, the injury from the latter may be expected to be
considerably greater than that from the former, on account
of its shape. The wedge-like quality of the conical bullet
is rendered particularly obvious on its being driven into the
shafts of the long bones of the extremities. The solid, os-

seous texture of which the cylindrical portion of these bones is composed is split up into fragments, having mainly a direction parallel with the central cavity; and fissures not unfrequently extend from the seat of injury to their terminanations in the joints, of which they form component parts. Such results were scarcely ever noticed from the impulse of round balls. The bone might be comminuted, but the fragments were of a more cuboid shape, and the long fissuring did not occur. It has been stated that the screw motion impressed on the ball by the rifling of the musket contributes to its increased power of injury on bone; but its shape, combined with its momentum, seem sufficient to explain the severity of its effects above those of the round bullet. Another result of the tapering form of the conical bullet is that it is less exposed, in its course through soft parts of the body, to opposition from tendons and other long and elastic structures, so frequently noticed to stay the progress of spherical shot. If not dividing them by direct impingement, it readily turns them aside; and it is partly due to this pointed shape, therefore, as well as to increased force, that, as will be noticed hereafter, the lodgment of balls is now so rare in comparison with the experience of former wars.

Much has been written on the comparative surgical effects of bullets of various weights and sizes; but these qualities do not, on consideration, excite so much practical interest in the mind of the surgeon as it might at first appear they are calculated to do. Some very heavy bullets were used by the Russians in the defense of Sebastopol, nearly one-third heavier than any employed by the troops opposed to them. Such bullets, if of like form and density, and propelled with equal velocity, would obviously inflict injuries—especially against osseous structures, which offer great resistance—wider in proportion to their greater size and momentum; but, in respect to simple flesh wounds, the increased size of the wound left by the larger ball would make little difference in the gravity of the wound, or the time required for its cure, while the escape of foreign substances, which it might happen to carry with it, would be facilitated by the freer means of exit and increased discharge from the surface. Mr. Guthrie mentions that, having had a wide field for observation in the effects of the heavy British musket-ball, sixteen to the pound, on the French wounded, he did not think them more mischievous in their results than the French musket-balls, twenty to the pound, on the English soldiers ; while the advantages of carrying a lighter musket and greater number of rounds of ammunition were on the side of our adversaries. It is understood that in warfare the object is not so much to destroy life as to disable antagonists, and the smaller size has been supposed to be fully equal to this object by the British military authorities of the present day; for in the weapon most recently given to the troops, the Enfield rifle, the weight of the ball has been reduced two drachms and a half below that of the ball with the Minié, previously in use. After all, within the moderate limits which must be preserved to suit the circumstances of infantry soldiers, the form and velocity of musket-balls must be the qualities of interest to the surgeon in connection with the wounds inflicted by them, rather than their weight or size, as with projectiles from guns of large caliber.

Double bullets, linked together by a spiral coil of wire, something after the manner of chain cannot-shot, were introduced by the Russians during the war in the Crimea. Specimens of these bullets were found about the works around Sebastopol, but no injuries received from them have been recorded ; although, after the discovery, peculiarities in the characters of some wounds, which had not previously been satisfactorily accounted for, were supposed to have probably resulted from them. It seems likely, however, that, when discharged, the divergent forces impressed on the two bullets were sufficiently great to break apart the connecting wire, which was of very slender diameter, before they came into contact with the troops against whom they were directed.

Dr. Scrive, in his History of the Eastern Campaign, mentions also that incendiary balls were employed by the Russians. They consisted of a small cylinder of copper containing a detonating composition, and made up into the form of an ordinary cartridge, so as to be discharged from a musket. On hitting its object, the projectile burst with violence. These balls were not known till after the conclusion of the siege ; and it was only then, M. Scrive remarks, that a key was obtained to some wounds of a frightful character which could not be accounted for by the action of ordinary bullets or fragments of shell. No similar observation is recorded in the British surgical history of the war.

Wounds caused indirectly by stones from parapets, splinters of iron or wood, and by fragments of shells are very varied in character and severity. They derive their importance chiefly from the extent of surface usually lacerated and destroyed. Unless they happen to have penetrated or torn away largely the coverings of vital parts of the body, they are often less grave, though to the sight more fearful, than injuries of less alarming appearance from direct projectiles. In missiles of this secondary kind, the amount of resistance offered to their displacement proportionably diminishes the impetus with which they strike. In like manner, the powerful opposition of the hollow iron shell to the force of the bursting charge within, as well as the shape of the portions into which it is usually rent asunder, combine to cause the momentum of each fragment at starting to be much less, and this momentum to be more rapidly retarded during its flight through the air, than happens in ordinary missiles of direct explosion. The constitutional shock, in these injuries, is consequently, as a general rule, less than in direct gunshot wounds. Occasionally simple fractures happen from indirect missiles ; from direct, they are almost necessarily compound. Although there may be no communication with an external wound, however, there is often great comminution of the bone in these accidents. The laceration and bruising of the soft parts are frequently rendered more dangerous from in-direct projectiles in consequence of large vessels or nerves being implicated in the injuries, leading more often to pri-mary hemorrhage and subsequent sloughing of wider tracts than in wounds from direct projectiles of corresponding size. Such sloughing may lead to a fracture of bone becoming compound which was at first simple. Fragments of shells sometimes wound by falling, after having been projected up-ward in the air. These do not generally produce such seri-ous injuries as fragments striking at once from the exploded shell; not that the force is different, but because the parts chiefly exposed—the shoulders, back, etc.—are more protected from injury, and offer less resistance, from relative form and position, than do the abdomen, loins, and other parts of the body, which usually meet the fragments shot upward when the shell explodes on the ground.

Degree of velocity.—The velocity of motion of different projectiles is an important ingredient in the consideration of the several wounds produced by them. The rates of motion imparted to missiles by the fire-arms of early times were probably, from the imperfect construction of the weapons, defective quality of gunpowder, and other circumstances, as inferior to those of the musket lately in use as the velocity of musket-balls was to that of the conical bullets of the rifles in present use. In a table showing the velocities of certain moving bodies, published in 1851, the common musket-bullet is set down as moving at the rate of 850 miles per hour, the rifle-ball of that time at 1000, the 24-lb. cannon-ball at 1600 miles per hour. But the musket-ball then could not be de-pended on to hit an object beyond 80 yards, the rifle 200 to 250 yards; while the present Enfield rifle is sighted to 900 yards, and the short Enfield to 1100 yards. The effects of different rates of velocity on wounds are seen in the varia-tions which occur in proportion to the distance which the missile has traveled before inflicting the injury. A cannon-ball which, with but slight velocity of motion added to its weight, would knock a man over, at ordinary speed will carry away a limb without disturbing the general equilibrium of the body. A musket-ball that would be arrested half way through a limb is now replaced by a ball which, at like distance from the point of discharge, will pass through several bodies in succession.

The increased velocity, or, in other words, greater force, of modern projectiles exhibits its effects in two directions—locally, by the greater destruction of the tissues in the track of the projectile ; and constitutionally, by greater disturbance in the nerve-force of the whole system. The component parts of that portion of the organized fabric through which a bullet, traveling at the rate of several miles per minute, cleaves its way are inevitably deprived of their vitality. Instances are quoted by authors, of gunshot wounds having healed by simple adhesion ; but such examples are not met with from rifle-bullets retaining their original form. Moreover, when considering the course taken by balls in the body, it will have to be shown that the velocity imparted to projectiles from modern weapons has led to another change in gunshot wounds. The great power of resistance so often before exhibited by the yielding elastic tissue of the skin, by tendinous and other structures, is no longer of avail against projectiles from modern fire-arms at their usual rates of speed.

The splitting and destructive effects of conical balls on the shafts of the long bones of the extremities have already been mentioned when referring to the peculiarities of their shape. But, together with form, the amount of momentum is a necessary ingredient in estimating this result. The old round balls—partly from their form, but also from the imperfect mechanism of the firelocks from which they were discharged, and consequent minor degree of velocity imparted to them—on striking bones, would simply be turned away from the direct line, or, failing this, would knock out a portion of the shaft without further fracture, or, having perforated on one side, remain in the cancellated structure, or be

3

simply flattened without penetrating. It seems not unlikely,
also, that the modern conical bullets are denser, from the circumstance of their manufacture by mechanical pressure, than
bullets, such as are still used in some places, cast in moulds.
The influence of density with respect to power of penetration is very great. In the two most perfect of modern English rifles, the Enfield and the Whitworth, the projectiles and
charges being of the same weight, when lead is used, the penetration at 800 yards is one-third greater with the Whitworth
than the Enfield ; but if a less yielding projectile is used, (as
when the lead is mixed with tin,) its penetration is as 17 to
4 at 800 yards. Whether this cause operates or not, the fact
is certain that conical balls in action exhibit almost invariably an overpowering force over all the structures, bone included, with which they come into contact in the human body,
and are rarely met with flattened, or so much altered in
form as bullets not unfrequently were formerly under like
circumstances.

Number of wounds in battle.—The increased velocity
of modern projectiles, together with the more rectilinear path
in which they move, causes a greater number of wounds in
modern warfare. The difference which has existed in the
proportion of wounded to shots discharged in recent engagements, compared with the experience of former wars, is
most marked. It is well known that from expansion of the
bore of the musket in use a few years since, and consequent
increase in the difference between its diameter and that of
the bullet, after a few rounds of fire musket-balls rolled out
in numerous instances in the act of elevation of the musket
previous to discharge. Now every shot is propelled to a
great distance, and with force sufficient, if brought into collision early in its flight, to penetrate and wound several persons. Colonel Wilford, Chief Instructor at the Government
School of Musketry, stated publicly in a recent lecture the
fact that 80,000 rounds of ball-cartridge were fired from the
old musket in one day in Caffraria, and only 25 Caffres were

known to be hit; while at Cawnpore, one company of soldiers, armed with the Enfield rifle, brought down 69 out of
a body of horsemen by whom they were attacked, at one discharge. At the battle of Salamanca, only one ball in 3000
fired by the British took effect. Another result is, that we
may now expect to meet more frequently the occurrence of
several bullet wounds in the same individual. It is mentioned that, among the wounded from Solferino, it was not
uncommon to see several wounds of different origins in one
body; and M. Appia mentions a case, in one of the hospitals at Brescia, where a soldier had been struck at the same
time by four balls. These circumstances become important
in estimating the amount of surgical attendance that is required in modern engagements. At the battle of Solferino,
just referred to, some returns show that, in twenty-four hours,
11,500 French, 5300 Sardinians, and 21,000 Austrians were
laid *hors de combat*. The surgeons had no time to attend
to the first necessities of a great proportion of the wounded.
A multitude of those unfortunates were hastily conveyed to
the little village of Castiglione, and had to wait hours, many
even days, before their wounds could be dressed. To relieve
thirst, and apply wet compresses of linen to ease the pain of
the wounds, by calling into service the people of the neighborhood, was as much as could be done to a great number
for the first day or two, on account of the vast number of
wounds inflicted by the new weapons. At Brescia, within
a short time after this battle, 15,000 wounded were congregated in thirty-eight fixed and temporary hospitals. From
the actions in Flanders on the 16th, 17th, and 18th of June,
1815, including the battles of Quatre Bras and Waterloo, the
returns show the number of wounded, not including those
killed in action, in the Duke of Wellington's army, to have
been rather more than 8000. In the whole Crimean campaign, the total number of British wounded amounted to
11,361, exclusive of men killed in action.

Spent balls.—In connection with degree of velocity, the

subject of what are called "spent balls" naturally occurs. After, a cannon-ball has ceased to pursue its course through the
air or to proceed by ricochet, it not unfrequently travels to a
considerable distance, rolling along the surface of the ground.
When its rate of movement is not much faster than that at
which a man can walk, and when to all appearance it might
be stopped by the pressure of the foot as readily as a cricket-
ball, it yet possesses the power of inflicting serious injury on
such an attempt being put into execution. This power is
easily understood if the amount of force is remembered which
must still be inherent in the cannon-ball for it to overcome
the inertia of its own mass, and the resistance to which it is
exposed in passing over the ground on which it is rolling.
It is this force, multiplied by the weight of the ball, which
gives it the destructive power. If this ball is brought into
collision with the foot of a person, such destruction ensues
as generally to necessitate amputation. Should it impinge
on other parts of the body, as in the instance of a man lying
on the ground, it may cause mortal injury to internal organs,
and that without exhibiting external evidence of the amount
of injury it has inflicted. So, also, though powerless to
carry away a limb, it may cause comminuted fractures of
bones and extensive contusions of the softer structures.

Lodgment of balls.—Low rate of velocity leads to musket and other balls lodging in various parts of the body.
When the smooth-bore musket was in common use, lodgment
of balls was of frequent occurrence. In the first place, from
absence of sufficient initial velocity to effect its passage out
of the body, and, secondly, from its liability to be diverted
from a direct line, a round ball might be arrested in its
progress at any distance from its point of entrance. Conical balls lodge when their velocity has become nearly expended before entering the body; or, from peculiarity in the
posture of the person wounded, a ball, having had force
enough to traverse a limb, may afterward enter into another
part of the body and lodge. A ball may reach a part so

deep in the muscles of the back, for example, or be so far removed from the aperture of entrance, as to elude all attempts on the part of the surgeon, at the time of examination of the wound, to discover its retreat. Or it may have reached some position from which the surgeon fears to take the necessary steps for its extrication, judging the additional injury that would thus be inflicted more mischievous than the probable effects of allowing the ball to remain lodged.

Unextracted balls lead to consequences varying according to the site of lodgment and state of constitution of the patient. If the ball have become fixed in the body of a muscle, or in its cellular connections, adhesive inflammation may be established around it, and in time a dense sac be thus formed, in which the ball may remain without causing any, or but very slight, inconvenience. M. Baudens asserts that a cellular envelope is of very early formation around balls lodged in muscular tissues. Although thus encysted, a ball may press upon nerves, and give rise to pain and much uneasiness, or may be so situated as to embarrass the person in certain movements of the body. Foreign bodies not unfrequently change the position of their first lodgment, under the effect of gravitation or the impulse of muscular actions. The following instance, which occurred to Staff-Surgeon Dr. Daniell, illustrates the distance to which a lodged ball may travel before finding its exit: In the disastrous affair of Malageah, on the west coast of Africa, fought in May, 1855, between detachments of the West India regiments and the Moriah chiefs, a man was wounded just below the spine of the scapula by a shot fired down from an elevation. The aperture was small, no ball could be traced, and the wound healed up rapidly. Six months afterward the man attended hospital, complaining of inability to march and pain about one of his ankles. A red, painful swelling and abscess formed over the inner malleolus, disease of bone was suspected, when examination led to the discovery of a small iron ball, of ir-

3*

regular shape, which was removed. No pain or irritation had existed between the shoulder and the foot. When lodged in the lower extremities, balls sometimes form for themselves canal-shaped cysts, along which they can be moved freely on pressure. When, however, the health or other circumstances of the patient are not favorable, the lodgment of a ball with a smooth surface, like missiles of a more angular and irregular shape, may excite inflammation and constitutional disturbance of a very troublesome kind, and keep up a profuse suppurating discharge along the track of the wound, or perhaps lead to abscesses burrowing in other directions. Balls have been known to lodge in bones, without their positions having been suspected or inconvenience excited by their presence. On the other hand, balls similarly impacted have given rise to disease, and in some bones, as those of the pelvis, have produced such constitutional irritation as to lead to a fatal termination. Balls lodging in the circumscribed cavities of the body or their contained viscera require notice elsewhere.

Grape-shot, and even balls of larger size from field guns, occasionally lodge. The large, gaping wounds inflicted by such missiles usually render the detection of their lodgment and position very easy; but still remarkable instances have occurred where the presence of bodies of this nature of very large size has been overlooked. Mr. Guthrie's experience of the war in the Peninsula led him to record that "it was by no means uncommon for such missiles as a grape-shot to lodge wholly unknown to the patient, and to be discovered by the surgeon at a subsequent period, when much time had been lost and misery endured." The same distinguished surgeon mentions a case where a ball weighing eight pounds was not discovered till the operation of amputating the thigh in which it had lodged was being performed. Baron Larrey describes a similar case : An artilleryman had his femur fractured by a ball, which, according to the man's description, had afterward struck another artilleryman by his side. On being brought to hospital, no one doubted that the ball, after fracturing the limb, had glanced off; but on amputating, the ball, weighing five pounds, was found in the hollow of the thigh toward the groin.   The wound of entrance was on the outside of the thigh; and the ball had not only fractured, but had turned round, the bone.   M. Armand, surgeon attached to the French Imperial Guard, has related the case of a soldier who was brought to the ambulance, after the taking of the Mamelon Vert, in the Crimea, with his left thigh wounded; one opening, such as might be made by a large musket-ball, was found on the outside of the thigh.   There was no second opening.   On examination, a swelling was detected in the popliteal space, without any external mark of injury nor much pain on pressure.   It was concluded to be the ball; and, on incising, an enormous grape-shot was found.   It had turned round the femur without breaking it.   M. Armand writes that the appearance of the wound alone would have led to the supposition that the ball had not lodged, and no one would have suspected that such a thing as a grape-shot had been the cause of it.   In the British Surgical History of the Crimean War the case of a soldier of the 1st Royals, who was wounded in the face by a grape-shot weighing 1 lb. 2 oz. is recorded.   The ball lodged at the back of the pharynx, and escaped observation for three weeks.   Were it not for experience of many such instances, it would be deemed almost impossible that foreign substances of such size and weight could remain in the body without the knowledge of the patient, if not discovered by the surgeon.   Even with so large a missile as a grape-shot, a surgeon should not be contented with examining merely by the wound, wide as it usually is, in case lodgment is suspected; it may travel in a direction which may cause its discovery to be very difficult by that track.   An officer of the 19th Regiment was struck during the assault on the Redan, on September the 8th, by two grape-shot, at the back of the chest.   They entered close to the spine.   One of these balls lodged in the inner part of the right arm, below the axilla, whence the writer excised it.

Penetrating fragments of shells, if projected edgeways, almost invariably lodge. In these cases, the appearance of the wound seldom indicates to the observer the true size of the body which has caused the injury. At an early period of the battle of the Alma, a piece of shell, about four pounds in weight, lodged in the buttock of a soldier of the 19th Regiment; and, to extract it, an incision had to be made nearly equal in extent to the length of the original wound. In this instance the concave aspect of the fragment—evidently, by the nature of the curve and thickness, a portion of a very large shell—had adapted itself to the parts lying beneath, while its convex surface so agreed with the natural roundness of the parts above, that it would have been impossible to have arrived at a knowledge of its lodgment, from any change in the external appearance of the parts. Examination by the wound alone gave decided information on the question. Such fragments become very firmly impacted among the fibers of the tissues in which they lodge, and the effused blood fills up inequalities, and rounds off edges that might otherwise 'show themselves prominently; so that, without due care, their presence is not unlikely to be overlooked at first examination. Dr. Macleod, of Glasgow, mentions that he saw a case at Scutari, in which a piece of shell weighing nearly three pounds was extracted from the hip of a man wounded at the Alma, which had been overlooked for a couple of months, and to which but a small opening led.* But bodies of still more irregular form may lodge in this region, and escape notice. A soldier in a battery in the Crimea was wounded, during a heavy artillery fire, in the left hip. A twelvemonth afterward he was in the General Hospital at Chichester, with a narrow sinus, which allowed a probe to pass deeply among the gluteal

---

* Notes on the Surgery of the Crimean War, p. 104, J. B. Lippincott & Co.'s edition.

muscles. On cutting down in the direction indicated, a piece of stone was extracted, upwards of four ounces in weight. This man had passed through several hospitals before his arrival at Chichester.

Bullets scattered from canister or spherical case not unfrequently lodge; apparently in consequence of the direct velocity received from the primary discharge being disturbed, and lessened by the force of the secondary explosion of the case in which they were contained.

A small layer of metal, like a portion of one of the coats of an onion, occasionally becomes detached from a leaden bullet, and lodges. The writer was once applied to by a discharged soldier, suffering from some troublesome granulations at the bottom of the left orbit. The globe of the eye had been destroyed nearly two years before by a musket-ball shot from above, which, after traversing the orbit, had descended, and was excised from the right side of the neck. On examining the granulations by a probe, the point came into contact with a hard substance, which further examination showed to be a small projecting point of lead. It proved to be a scale from the bullet which had caused the original wound, being equal in length to half its circumference, and in width, at the broadest part, about a third of the same dimension. It retained the curved form of the bullet from which it had been detached. The following case shows that similar sections may be separated from cylindro-conical as well as from round bullets. An officer of the 41st Regiment was struck in the Crimea by a conical bullet, which destroyed the forearm in such a manner as to necessitate amputation below the elbow. Secondary hemorrhage occurred on the eleventh day, and on the following day the stump was opened and examined. "While searching for the bleeding vessel, a slice of the bullet, about the size of a worn sixpence, was found deeply imbedded in the muscle." In the case of a soldier of the 19th Regiment, who was wounded before Sebastopol in the loin by a conical bullet,

which was discharged per anum, and who died in Guy's
Hospital of albuminuria, nearly four years afterward, a small
scale of lead from the bullet was found at the post-mortem
examination fixed in the spleen. Strange to say, in this instance the lodgment did not appear to have excited any
inflammatory action or mischief.

Lodgment of small foreign bodies, angular pieces of metal,
as slugs, nails, and others, and of soft textures, as shreds of
linen or woolen cloth, often give rise to much inconvenience.
The track of a musket-ball may be prevented from healing, and a troublesome sinus formed, by such small fibers of
cloth as would hardly attract notice if within means of observation. Although a wound be closed, and apparently
healed, if any shreds of cloth remain, it will probably open
from time to time, when small fibers may be noticed in the
discharge; and this will continue until the whole is thus got
rid of. The probability of cloth entering a wound with the
conical ball is not so great as it was with the spherical ball,
which not unfrequently tore out a little cap, as it were, of
cloth in its passage. This is another result of its shape and
velocity. John Hunter and others make mention of circular
pieces of the skin being cut out by bullets, and then lodging,
and acting as foreign bodies in the wounds.

When the Minié-ball, with the iron cup at its base, was
first brought into use, surgeons anticipated that the addition
of the iron cup would complicate the ill effects of the wounds
inflicted by it. It does not appear that this has proved to
be the case. The iron is usually so far driven into the lead
by the force of the exploded gunpowder, and so firmly fixed
by the alteration in shape and pressure of the lower part of
the ball, that it but rarely becomes detached so as to form a
separate lodgment.

Gravel and small stones struck up by shells at the time of
their explosion, or by shot ricochetting against the ground,
often lodge, and give much trouble in their extraction, especially about the face. In the assault of Sebastopol, at the

Great Redan, the attacking parties in their approach, the ground being rocky and having been much broken up by shell explosions, were particularly exposed to such injuries; and in several instances men were placed *hors de combat* by dust and small fragments of stone thus projected, though the injuries were not of a permanently serious character. One case is recorded where both eyes were penetrated and totally destroyed by gravel thrown up by a shell explosion.

Foreign substances derived from persons standing near a wounded man, sometimes fragments of the bodies of other wounded men, have been already named as occasionally lodging. In a severe injury to the face, which occurred in a man of the 1st brigade of the Light Division, in the Crimea, the surgeon was at first puzzled by the strange displacement of a part of the upper jaw. After closer examination, and obtaining a clearer view by the removal of clot, it was found that a piece of the jaw of another man, whose head had been smashed by a round shot by his side in the battery, had been driven into the palate, and was there impacted. Among other cases recorded in the Surgical History of the Crimean War, is one of a double tooth of a comrade having been found imbedded in the globe of the eye; and another, where a portion of a comrade's skull was removed from between the eyelids of a soldier. In such injuries as these, where one of two men standing side by side is wounded by a portion of the body of his neighbor, the fragment striking is usually detached from a corresponding region with that struck. The late Mr. Guthrie extracted from the thigh of a Hanoverian soldier, on the third day after his admission into hospital, two five-franc pieces and a copper coin. The man had had no money about him previously to the injury, nor pocket to contain any. The coins had been carried from the pocket of a neighbor, who stood before him in the ranks, and who had been hit by the same grape-shot. These coins, flattened out and jammed together by the force of the shot, are in the museum at Fort Pitt. Similar examples might

be multiplied; but sufficient have been mentioned to show
the necessity of careful examination in warfare, not only for
direct missiles which may effect lodgment in the body, but
for many other foreign substances which may be forced in by
their agency.

Internal wounds without external marks.—Among
the wide variety of injuries from gunshot, there have not
unfrequently been noticed cases in which serious internal
mischief has been inflicted, without any external marks of
violence to indicate its having resulted from the stroke of a
projectile. An important viscus of the abdomen has been
ruptured, yet no bruising of the parietes observable; symptoms of cerebral concussion have shown themselves, yet no
injury of the scalp to be detected. Even bones have been
comminuted without any wound of the integuments or appearance of injury. The records of the Crimean campaign
afforded not unfrequent examples of such wounds. Two
cases occur, in the French records, of fracture of the forearm without any external apparent lesion; in one the internal structures were reduced to a mass of pulp. The difficulty of reconciling the several facts noticed in such instances,
together with the vague descriptions by patients of their
sensations, led surgeons to seek an explanation for them in
the supposition that masses of metal projected with great
velocity through the air might inflict such injuries indirectly
by aerial percussion. Either the air might be forcibly driven
against the part injured by the power and pressure of the
ball in its flight, or a momentary vacuum might be created,
and the forcible rush of air to refill this blank might be the
origin of the hurt. Electricity has also been called into aid
in explaining these injuries. All these hypotheses are now
abandoned. So many observations have been made of cannon-balls passing close to various parts of the body, as near
as conceivable without actual contact, without any such consequences as those attributed to windage, as to lead to the
necessary conclusion that the theory must in all instances

have been fallacious. Portions of uniform and accouterments have been torn away by cannon-balls without injury to the soldier himself. Even hair from the head has been shaved off, and cases are on record where the external ear and end of the nose have been carried away without further mischief.

The true explanation of the appearances presented in those cases which were formerly called "wind contusions," appears to rest in the peculiar direction, the degree of obliquity, with which the missile impinges on the elastic skin, together with the situation of the structures injured beneath the surface, relatively to the weight and momentum of the ball on one side, and hard resisting substances on the other. Thus, in the case of a cannon-ball passing across the abdomen, as in two instances mentioned by Sir Gilbert Blane, where men were killed by the passage of balls across the epigastrium, the elasticity of the skin probably enabled that structure to yield to the strain to which it was exposed, while viscera were ruptured by the projectile forcing them against the vertebral column. So the weight of a ball passing obliquely over a forearm may possibly crush the bone between itself and some hard substance against which the arm may be accidentally resting, without lesion of the interposed skin. Baron Larrey, who examined many fatal cases of this kind, relates that he always found so much internal disorganization as to leave no doubt in his mind of its being the result of contact with the ball. He explained the absence of superficial lesion, by the surface having been struck by cannon-balls in the latter part of their flight, when they had undergone a change of direction from straight to curvilinear, and acquired a revolving motion, owing to atmospheric resistance and the effect of gravitation. In such a condition, he argued, they would turn round a part of the body, as a wheel passes over a limb, in place of forcing their way through it; and, while elastic structures would

4

yield, bones and muscles, offering more opposition, would be bruised or broken.

In some recently published letters on the wounded in the late campaign in Italy, by M. Appia, this writer states that wounds from massive projectiles having been rare, he had not met with an example of internal destruction of parts with skin preserved intact, and that he had nowhere seen a wound which was attributed to *vent de boulet*. The hypothesis, he remarks, seems generally abandoned. It is presumed that, in stating wounds from *gros projectiles* to have been rare, he refers only to the wounded in the hospitals, and that it is to be inferred that the injuries from cannon-shot proved generally fatal in the field.

Seat of injury.—A knowledge of the seat of injury from the passage of a ball involves diagnosis of its course, the depth of its penetration, the particular organs or structures injured, and the extent of the injuries to which they have severally been subjected. The course pursued by balls in wounds presents many features of interest. The depth of penetration, in connection with direction, becomes of great importance when there is question of one of the great visceral cavities being opened. This part of the subject, however, together with that of injuries to the viscera themselves, will be more conveniently considered when treating of gunshot wounds in their special relations to particular regions. In like manner, the diagnosis of the extent of injury in wounds complicated with fractures of the long bones will be best considered under gunshot wounds of the extremities.

Course of balls.—Of the circuitous and unexpected directions pursued by bullets in their course through the human frame, which were formerly so common, we are not likely to see many instances in future warfare, when the rifle is the weapon chiefly employed. The conical shape of the ball and the force with which it is propelled have had the effect, among others already named, of changing this characteristic of the ball from the smooth-bored musket. The latter, bearing a force that scarcely carried it true to a mark at eighty yards, and usually receiving, as it left the firelock, an impulse which caused it to revolve on its axis at right angles with the line of flight, was deflected by the most trifling obliquity of surface, by the resisting obstacle of a bone, by tendons or the aponeuroses of muscles, or even by the elastic resilience of muscles themselves in a state of action, when the relative direction of their fibers was favorably placed to exert this influence. The Enfield cylindro-conoidal bullet, armed with a force that will carry it to a given spot distant one thousand yards or upwards, flies like an arrow, penetrates the softer tissues in a straight line, and on meeting bone, as before noticed, enters it like a wedge. When a bullet of this kind strikes an object point-blank, it is always the apex of the conical part which first meets the object struck; and, if sufficient resistance be met with, it is this apex which becomes first compressed and turned back. When it strikes a solid object lying nearly parallel with its line of flight, the ball is planed, as it were, from its apex toward its base. In a case before referred to—page 29—where a conical ball entered the loin of a soldier of the 19th Regiment, and was subsequently passed per anum, the apex of the bullet was found to be turned and bent round on itself, and the ball generally flattened. On examining carefully the convex surface of the convoluted apex, minute spiculæ of bone were observed to be impacted in its substance. It became evident, therefore, that the ball had struck, probably penetrated through, some portion of the lumbar vertebræ in its course from the loin to the intestine. There were no general symptoms to indicate spinal injury, but, four years afterward, the opportunity of a post-mortem examination being afforded, the track of the ball through some of the lumbar vertebræ was distinctly traced.*

---

* See Guy's Hospital Reports, 3d series, vol. v., 1859—case of Gunshot Wound in the Loins, by S. O. Habershon, M.D.

It will often appear, at first examination, that the track of a wound by the cylindro-conoidal bullet, even at full speed, is widely removed from a straight line, especially when this class of injuries is new to the surgeon. It is not difficult to understand the apparent irregularity in the line of the wound, when the many varied positions in which the body and its parts are liable to be placed are called to mind, and if, when making the examination, the surgeon has omitted to place the patient in a similar posture to that he was in when struck. A certain allowance must also be made for the spasmodic actions of the various muscles among themselves, and momentary displacement of other structures, at the instant of receiving the injury.

Occasionally, though rarely, an accidental concurrence of circumstances may lead to the conical bullet pursuing a circuitous instead of a direct course, especially when, after traveling a certain distance, its speed has become diminished; and, as round musket-balls are not yet wholly discarded from warfare, it is necessary to call attention to the observations which have been made on this subject. Balls have been known to pass round the outer convex and the inner concave surfaces of the abdominal and thoracic cavities, sometimes forcing their exit at points nearly opposite to those of entrance, sometimes making a complete circuit. Thus, from simple observation of the line of direction of two wounds, a ball may be supposed to have passed through the thorax or abdomen, while really it may not have penetrated the cavity, but only made its way beneath the integument. In like manner, a lung may be supposed to have been traversed by a ball, not merely from the relative position of the wounds of entrance and exit, but also by some of the characteristic signs of such an injury, when really the ball, after entering the cavity of the chest, has rolled round the costal pleura, never penetrating the lung, but at the most bruising its surface. In the same way, balls have been known to travel round the cranium beneath the scalp, and to have found their way beneath the integuments of the neck, without injury to the deeper structures. Dr. Hennen saw a case where a ball was found lying in a wound by the thyroid cartilage. It had made a complete circuit of the neck, and returned to the spot where it had entered. Cases sometimes occur where two openings are found in a man's shoulder, in such relation that a straight line between them would necessarily pass through the head of the humerus, yet the ball has only made a half circuit, outside the joint.

Many examples of such injuries will be found in the works of all writers on gunshot wounds until the recent introduction of rifled weapons, while those who have only seen the latter in use are almost inclined to doubt the accuracy of previous statements on this subject, from not meeting with similar instances in their own experience. In the early part of the late war with Russia, the musket wounds were nearly all inflicted by the round bullet; but during the year 1855 conical bullets of various shapes and sizes were brought into use by the Russians generally, as they had been for some time previously by nearly the whole of the English army, and a large proportion of the French army. As early as the battle of Inkerman, however, the Russians were partly armed with the Liège rifle, with its conical bullet. Among 3000 wounded from the recent battles of Palestro and Magenta, assembled in the hospitals at Turin, M. Appia, whose letters on the wounded in the late Italian campaign have been before quoted from, writes that he was astonished not to meet one case of a cylindrical ball having taken a curved direction in its passage. He mentions the case of an officer being wounded by a ball, which entered at the epigastrium and passed out by the side of one of the lumbar vertebræ, without penetration of the abdomena, red mark or zone connecting the two wounds and indicating the circuit which the ball had made. In another case, a ball had traversed the chest from right to left, and still had sufficient force

4*

to wound the left arm. Both these injuries, however, were caused by spherical balls.

SYMPTOMS OF GUNSHOT WOUNDS.

The leading symptoms of gunshot wounds are the diagnostic marks of these injuries, and the constitutional disturbance, pain, hemorrhage, edema, and other circumstances with which they are attended. Some of these require to be noticed separately.

Diagnosis.—The external distinguishing signs of a penetrating gunshot wound are generally manifest enough, but exact diagnosis of the nature and extent of the wound is not always so simple as it might at first appear to be. It is necessary to describe, firstly, the external appearances. These, although possessing certain universal characteristics, vary to a wide extent, according to the different forms, already described, of the missiles causing the injuries, their velocity, the part of the body struck, and its position relative to the projectile at the time of injury.

When a cannon-ball at full speed strikes in direct line a part of the body, it carries away all before it. If the head, chest, or abdomen are exposed to the shot, an opening corresponding with the size of the ball is effected, the contiguous viscera are scattered, and life is at once extinguished. If it be part of one of the extremities which is thus removed, the end remaining attached to the body presents a stump with nearly a level surface of darkly contused, almost pulpified, tissues. The skin and muscles do not retract, as they would had they been divided by incision. Minute particles of bone will be found among the soft tissues on one side, but the portion of the shaft of the bone remaining *in situ* is probably entire.

In ricochet firing, or in any case where the force of the cannon-shot is partly expended, the extremity, or portion of the trunk, may be equally carried away, but the laceration of the remaining parts of the body will be greater. The surface of the wound will be less even. Muscles will be separated from each other, and hang loosely, offering at their divided ends little appearance of vitality; spiculæ of bone of larger size will probably be found among them; and the shaft may be found shattered and split far above the line of its transverse division. The injury to nerves and vessels may be proportionally higher and greater. Occasionally it happens, even where the limb seems to have been struck in direct line, that it is nevertheless not completely detached, but remains connected by shreds of the skin and parts of the tissues, on which the bone, reduced to minute fragments, is mixed with the contused muscles and other soft parts in a shapeless mass.

If the speed be still further diminished, so that the projectile becomes what is termed a "spent ball," there will not be removal of the part of the body struck, but the external appearances will be limited usually to ecchymosis and tumefaction, without division of surface; or even these may be wanting, notwithstanding the existence of serious internal disorganization. The rationale of such phenomena has been previously described.

Should the cannon-ball strike in a slanting direction, the external appearances of the wound will be similar to those just described, according to its velocity, modified only in extent by the degree of obliquity with which the shot is carried into contact with the trunk or extremity wounded.

Large fragments of heavy shells generally produce immense laceration and separation of the parts against which they strike, but do not carry away or grind, as round shot. Ordinarily, the line of direction in which they move forms an obtuse angle with the part of the body wounded. When they happen to strike in a more direct line, so as to penetrate, the external wound, as alluded to under the head of lodgment of projectiles, is mostly much smaller than the fragment itself, from the projectile not having had force enough to destroy the vitality and elasticity of the soft parts through which it entered.

Small projectiles, with force enough to penetrate the body, leave one or more openings, the external appearances of which also vary according to their form and velocity. The appearance of a wound from a rifle-ball, at its highest rate of speed, may be sometimes witnessed in cases of suicide. A soldier, in thus destroying himself, mostly stoops over the muzzle of his firelock, pressing it against the upper part of his body, and springing the trigger by means of his foot. The muzzle is usually applied beneath the chin. In such a case, a circular hole, without any puckering or inversion of the marginal skin, together with dark discoloration of the integument for several inches round, is observed at the wound of entrance. The vertex of the head is shattered; fragments of the parietal and occipital bones, together with small portions of brain, are carried away and scattered about; the bones not broken are loosened from their sutures; the mass of brain is torn to pieces, but held by its membranes; the superficial vessels of the face are distended with blood. These effects are not wholly due to the passage of the ball, but partly to the flame from the ignited gunpowder jetting out at the mouth of the musket, and in part also to the expansive force exerted within the cavity of the cranium, by the gases resulting from the explosion.

When the musket-ball strikes at a distance from the weapon by which it was propelled, but still preserves great velocity, the appearances of the wound are changed. An opening is observed, irregularly circular, with edges generally a little torn; and the whole wound is slightly inverted. There may be darkening of the margin, of a livid purple tinge, from the effects of contusion, or it may be simply deadlike and pale. Should the ball have passed out, the wound of exit will be probably larger, more torn, with slight eversion of its edges and protrusion of the subcuta

neous fat, which is thus rendered visible. These appearances are the more easily recognized, the earlier the wound
is examined. They are more obvious if a round musket-ball
has caused the injury than when it has been inflicted by a
cylindro-conoidal bullet. Indeed, with the latter, where it
has simply passed through the soft tissues of an extremity
of the body at full speed, it is usually very difficult to dis-
tinguish by its appearance the wound of entrance from that
of exit. In medico-legal investigations concerning gunshot
wounds, it must be often a matter of great importance to
decide this point; but to the military surgeon, more especially from the circumstances connected with the new projectiles, it has become a subject of little practical interest.
When the indirect and tortuous penetration of balls was
the rule rather than the exception, a knowledge of the spot
at which the ball entered was often useful in diagnosing the
mischief it had probably committed in its passage, and in
determining the part of the wound where foreign bodies might
be supposed to be carried and to be lodging. When the track
of the ball is nearly in a straight line, as now usually happens, such information cannot be looked for from knowing
the relation of either opening to the entrance or passage of
the missile.

A musket-ball ordinarily causes either one wound, as when
after entering it lodges, or, as sometimes happens, from its
escaping again by the wound of entrance; or two wounds,
from making its exit at some point remote from the spot
where it entered; but occasionally leads to a greater number of openings. This last result may happen from the ball
splitting into two or more portions within the body, and
causing so many wounds of exit. A case occurred to M.
Dupuytren, where a ball split against the spine of the tibia;
and after traversing the calf of the leg in two directions,
entered the other leg at two points,—one ball thus causing
five orifices. A case occurred to the writer, in the Crimea,
where a cylindro-conoidal rifle-ball with three canalures,

after fracturing the cranium, was cut in two by the upper
edge of bone at the seat of fracture, smoothly as if by a
sharp instrument.   One part glanced off, the other entered
the cranium.   A strange feature in this case was, that the
depressed portion, after admitting the ball, closed up again;
so that no aperture, but only a slight depressed line of fracture, was visible.* A somewhat similar case occurred in the
38th Regiment, but the ball appears to have been a round
one. M. Huguier has collected some curious cases of splitting
of balls, from the records of the French revolution : among
others, the division of a ball into two parts, of another into
three parts, against the supra-orbitar ridge, and of another
into three parts against the clavicle.   A case is recorded,
where a grenadier in Algeria was wounded in five places, all
wounds of entrance, by one ball.   It was divided into five
portions by first striking against a rock at five or six paces
from the soldier, the fragments rebounding at various angles.
John Hunter mentions the case of a young gentleman who
was shot through the abdomen by means of a musket loaded
with three balls. In this instance there were only two orifices
of entrance and two of exit, one ball having followed in the
track of one of the others; "that there were three that went
through him was evident, for they afterward made three
holes in the wainscot behind him, but two very near each
other." Had it not been for this proof, it being known that
three balls were discharged, a suspicion might have existed
that one of the three balls had lodged.   The recollection
that such accidents may occur will sometimes assist in the
diagnosis of doubtful cases.

The number of wounds made by one ball may be increased
by its traversing two adjoining extremities of the same person, or even distant parts of the body from accidental rela-

* The portion of cranium referred to, with the piece of ball weighing half an ounce, which lodged in the cerebrum, are in the museum
at Fort Pitt.

tive position at the time of the injury.   On the 18th of June,
1848, at Paris, a man received a ball in his right arm, above
the elbow, which caused a comminuted fracture of the
humerus.   It then passed across and entered the left arm
below the elbow, fracturing the upper part of the radius.
Dr. Hennen mentions the case of a man on a scaling-ladder,
in which a ball passed from the middle of the upper arm on
one side to the middle of the thigh on the opposite side.   It
is evident, when the ball traverses with sufficient velocity,
that these accidents will not unfrequently occur, especially
between the upper extremity and trunk.   They correspond
with such events as more than one person being wounded
by the same ball, examples of which were not unfrequently
noticed in the trenches before Sebastopol, from enfilading
shots, especially prior to the capture of the Mamelon Vert
and other outworks; and are said to have been very common
in the late campaign in Italy.   Should the Whitworth rifle
ever be brought into general use, the proportionate number
of wounds thus caused from the greater density of the ball,
its immensely superior force, and low trajectory, must be
still further increased.

The two openings made by one ball may hold such a relative situation as to lead to the mistake of their being supposed to be caused by two distinct balls.   A case is recorded
where a ball entered the scrotum, and made its exit from
the right thigh, without any intermediate mark of its passage ; such a wound might lead to an erroneous diagnosis
of this sort.   Length of traverse, and consequent distance
between the two openings, parts of the body brought into
unusual relations from peculiarities of posture, and peculiar
deflections of the ball, may all be sources of this error.

The appearances of wounds resulting from penetrating
missiles of irregular forms, as small pieces of shells, musket-
balls flattened against stones, and others, differ from those
caused by ordinary bullets in being accompanied with more
laceration, according to their length and form.   Being

usually projected with considerably less force than direct missiles, such projectiles ordinarily lead only to one aperture, that of entrance.

Pain.—A gunshot wound by musket-shot is attended with an amount of pain which varies very much in degree according to the kind of wound, condition of mind, and state of constitution of the soldier at the time of its infliction. It will sometimes happen in simple flesh wounds, that patients will tell the surgeon they were not aware when they were struck; and examples attesting the truth of such statements occur, of soldiers continuing in action for some time without knowing they had been wounded. Sometimes the pain from the shot is described as a sudden smart stroke of a cane; in other instances as the shock of a heavy intense blow. Occasionally the pain will be referred to a part not involved in the track of the wound. Lieutenant M. of the 19th Regiment was wounded by a musket-ball at the assault of the Redan, on the 8th of September, 1855. His sensations led him to imagine that the upper part of his left arm was smashed, and he ran across the open space in front of the works, supporting the arm which he supposed to be broken. On arriving at the advanced trench, he asked for water; on trying to drink, he found that his mouth contained blood, and that he was unable to swallow. The arm, on examination, was found to be uninjured, but a ball had passed from right to left through his neck, and from its direction had no doubt struck some portion of the lower cervical or brachial plexus of nerves. Immediately after the transit of a ball, the sensibility of the track and parts adjoining is found to be partially numbed, so that examination is borne more readily for a short time after the accident than at any later period. Of course, after reaction sets in, or when inflammation has become established, the pain of the wound is proportionably increased. When a ball does not penetrate, but simply inflicts a contusion, the pain is described to be more severe than where an opening has been made by it.

Shock.—When a bone is shattered, a cavity penetrated, an important viscus wounded, a limb carried away by a round shot, pain is not so prominent a symptom as the general perturbation and alarm which supervene on the injury. This is generally described as the "shock" of a gunshot wound. The patient trembles and totters, is pale, complains of being faint, perhaps vomits. His features express anxiety and distress. This emotion is in great measure instinctive; it is witnessed in the horse hit mortally in action, no less than in his rider; it is sympathy of the whole frame with a part subjected to serious injury, expressed through the nervous system. Examples seem to show that it may occasionally be overpowered for a time, even in most severe injuries, by mental and nervous action of another kind; but this can rarely happen when the injury is a vital one. Panic may lead to similar results when the wound is of a less serious nature. A soldier, having his thoughts carried away from himself—his whole frame stimulated to the utmost height of excitement by the continued scenes and circumstances of the fight—when he feels himself wounded, is suddenly recalled to a sense of personal danger; and if he be seized with doubt whether his wound is mortal, depression as low as his excitement was high may immediately follow. This will happen according to individual character and intelligence, state of health, and other circumstances. For while, on the one hand, numerous examples occur in every action of men walking to the field hospital for assistance almost unsupported, and with comparatively little signs of distress, after the loss of an arm or other such severe injury; on the other, men whose wounds are slight in proportion are quite overcome, and require to be carried.

As a general rule, however, the graver the injury, the greater and more persistent is the amount of "shock." A rifle-bullet which splits up a long bone into many longitudinal fragments, inflicts a very much more serious injury than the ordinary fracture effected by the ball from a smooth-bore

5

musket, and the constitutional shock bears like proportion.
When a portion of one or of both lower extremities is carried away by a cannon-ball, the higher toward the trunk the
injury is inflicted, the greater the shock, independent of the
loss of blood. Some writers, in accounting for "shock,"
lay stress on the concussion, and general mechanical effects
on the whole body, of the momentum of the iron shot.* To
a certain extent this may be true, but, judging from analogy
in physics, the greater the velocity, and consequently the
momentum, of a ball carrying away a limb, the less would
the concussion of the trunk and distal parts of the body be.
A pistol-ball at full speed will take a circular portion out of
a pane of glass without disturbing the remainder; if the
speed be much slackened, as when fired from a distance, it .
will shake the whole pane to pieces.

That true "shock," (*ébranlement* of French writers,) as
distinguished from shock resulting from mental depression
after unusual excitement, or the effects of groundless alarm
on the part of a patient, is a phenomenon the essential rela-

---

* In the Medical and Surgical History of the War against Russia
in the Years 1854–55–56, published by authority, vol. ii. p. 265, the
physical effects of concussion in producing "shock" are strongly
dwelt upon. It is remarked: "The shock of the accidents frequently
witnessed by the military surgeon differs, often in a very material
degree, and possibly in kind also, from that witnessed in civil life.
When a cannon-shot strikes a limb and carries it away, the immense
velocity and momentum of the impinging force can scarcely be supposed to have no physical effect upon the neighboring or even distant
parts independent of, and in addition to, the 'shock,' in the ordinary
acceptation of the term, which would result from the removal of the
same part by the knife of the surgeon, or the crushing of it by a
heavy stone or the wheel of a railway wagon.   *   *   In the great
majority of cases, the whole frame is likewise violently shaken and
contused, and, probably, independent of these physical effects, a further vital influence is exerted, which exists in a very minor degree,
if at all, in the last-named injuries, and may possibly depend upon
the ganglionic nervous system."

tions of which are connected with vital force, and with that endowment of the organization only, may be judged from observation of cases in which the direct result of the wound is inevitably fatal, including many where no physical effects on neighboring parts from concussion could possibly be produced. In such injuries the "shock" remains, from the time of first production of the fatal impression till life is extinguished. And the practical experience of every army surgeon teaches him that where a ball has entered the body, though its course be not otherwise indicated, the continuance of shock is a sufficient evidence that some organ essential to life has been implicated in the injury. That the shock of a severe gunshot wound may be complicated with other symptoms, or that some of its own symptoms may be exaggerated from other causes,—hopes disappointed, the approach of death, and all the attendant mental emotions,— scarcely affects the question at issue ; for its existence, independent of these complications, in all such cases is undoubted.

Primary hemorrhage.—Primary hemorrhage of a serious nature from gunshot wounds does not often come within the sphere of the surgeon's observation. If hemorrhage occur from one of the main arteries, it probably proves rapidly fatal ; and surgeons, after an action, are usually too much occupied with the urgent necessities of the living wounded to spare time for examining the wounds of the dead, who are mostly buried on the field where they fall. Thus most surgeons speak of primary hemorrhage being exceedingly rare, more rare, perhaps, than it actually is. M. Baudens, referring to his service in Algeria, has remarked that he has often found on the field of battle wounded soldiers who had died of primary hemorrhage.

In those wounds to which the surgeon's care is called, the primary hemorrhage is ordinarily small in quantity and of short duration—a sudden flow at the moment of injury, and nothing more. When a part of the body is carried away by

round shot or shell, the arteries are observed to be nearly in
the same state as they are found to be in when a limb is torn
off by machinery. The lacerated ends of the middle and
inner coats are retracted within the outer cellular coat; the
caliber of the vessel is diminished, and tapers to a point near
the line of division; it becomes plugged within by coagulum;
and the cellulo-fibrous investing sheath, and the clot which
combines with it, form on the outside an additional support
and restraint against hemorrhage. When large arteries are
torn across, and their hemorrhage thus spontaneously prevented, they are seldom withdrawn so far but that their ends
may be seen protruding and pulsating among the mass of
injured structures; yet, though the impulse may appear very
powerful, further hemorrhage is rarely met with from such
wounds. There is more danger of continued hemorrhage
from wounds by pieces of shell, as the arteries are liable to
be wounded without complete transverse section of their
coats. The sharp edges, less velocity, and oblique direction in which the fragments usually impinge sufficiently
explain this difference.

It comparatively rarely happens that arteries are cut across
by musket-bullets, either round or conical. The lax cellular
connections of these vessels, the smallness of their diameters
in comparison with their length, the elasticity as well as
toughness of the tissues forming their coats, the fluidity of
their contents, and, in consequence of all these conditions,
the extreme readiness with which they slip aside under
pressure, act as means of preservation when these important
structures are subjected to such danger as the passage of
a musket-ball in their direction. Endless examples occur
where the ball appears to have passed through in the direct
line of the artery, so that it must have been pushed aside by
it to have escaped division. Mr. Guthrie mentions a case
where a ball even opened the sheath of the femoral vessels,
and passed between the artery and vein, in a soldier at Tou-
louse, without destroying the substance of either vessel. So

close was the ball, and such contusion was produced, together with, doubtless, injury to the vasa vasorum, that the artery became plugged with coagulum, and obliterated. A preparation of these vessels is in the museum at Fort Pitt. Another case is mentioned by Mr. Guthrie, where the direction of a ball between the left clavicle and first rib, and permanent diminution of the pulse in the arm on the same side, led to the conclusion that the subclavian had escaped direct destruction by the missile in a similar way.

Vessels do not always thus happily elude division by the ball. Captain V., of the 97th Regiment, whose death led to so much interest in England, was struck by a ball which divided the axillary artery on the right side. The arm had apparently been extended when he received the injury, as if in the act of holding up his sword. The night was very dark, the distance from the place where the sortie took place in which he was wounded to the camp hospital was more than a mile and a half, and he sunk from hemorrhage while being carried up. The death of an officer from division of the femoral artery is recorded in the Surgical History of the Crimean War, where also cases are mentioned, though not immediately fatal, of a wound of the femoral vein and profunda artery in the same subject from a conical bullet; and another, of the popliteal artery and vein, also from a rifle-ball. Mr. Guthrie mentions the cases of two officers who were killed, almost instantaneously, one by direct division of the common iliac artery, the other of the carotid. Primary but indirect hemorrhage, in consequence of a gunshot injury, usually occurs as a complication of fractured long bones, the sharp points and edges of which, extensively torn up as they now are by conical bullets, are well calculated to cause such injuries. They are not as frequent as might be expected, from the limits within which the dispersion of the fragments is restricted by their periosteal and other connections, and the yielding mobility, before mentioned, of the vessels themselves. We have no data, how-

5*.

ever, to guide us in determining the proportionate frequency
of fatal results from primary hemorrhage after wounds; nor
can we have them until proper examination and classifica-
tion of the particular causes of death on the field of battle
are instituted.

PROGNOSIS.

Gunshot wounds vary in gravity from the simplest laceration of cuticle to the instantaneous destruction of life. Death
may take place primarily from direct causes already alluded
to, viz.: from the destruction of vital organs, from extreme
shock to the vital forces through the nervous system, or from
hemorrhage; or it may ensue indirectly from secondary hemorrhage, gangrene, erysipelas, hectic fever, pyemia, or from
the results of operations necessarily required in consequence
of the original injury. In estimating the probable issue of
a particular wound, not only the state of health at the time,
but, if a soldier, the previous service, and diseases under
which he has labored during it, must be taken into account,
and the circumstances in which he is placed with respect to
opportunity of proper care and treatment must also be carefully weighed. The time which has elapsed after the receipt
of the injury is another important matter in forming a prognosis. The difficulties which have been already enumerated
in the way of arriving at a safe diagnosis of the true nature
and extent of the injury, and the liabilities above mentioned
to which a patient with a gunshot wound is exposed, should
put a surgeon on his guard against giving a hasty judgment
in any case that is not very plain and simple. Military surgery abounds with examples of wounds of such extent and
gravity as apparently to warrant the most unfavorable prognosis, which have nevertheless terminated in cure; while
others, regarded as proportionably trifling, have led to fatal
results. Tables may be found in works showing statistically
the nature and relative numbers of wounds and injuries re-

ceived in various actions, with their immediate and remote
consequences, as well as the results of the surgical operations they have led to; but these afford little aid toward the
prognosis of particular cases, each of which must be esti-
mated in its own individual circumstances. Such tables are
chiefly of value where they afford indications of the effects of
different modes of treatment in wounds of a corresponding
nature, and then only in patients under like circumstances of
age and condition. Even moral circumstances must not be
disregarded. The probable issue in any given case will be
very different in one soldier, who is supported by the stimulating reflection that he has received his wound in a combat
which has been attended with victory, from what it will be
in another, who labors under the depression consequent upon
the circumstances of defeat.

TREATMENT OF GUNSHOT WOUNDS IN GENERAL.

When the circumstances of a battle admit of the arrangement, the wounded should receive surgical attention prelim-
inary to their being transported to the regimental or general
field hospitals in rear. A slight provisional dressing, a few
judicious directions to the bearers, may occasionally prevent
the occurrence of fatal hemorrage, or avert serious aggrava-
tion of the original injury from malposition, shaking, and
spasmodic muscular action, in the course of conveyance from
the neighborhood of the scene of conflict to the hospital.
In the siege operations before Sebastopol, this was accomplished by assistant surgeons in the trenches, or, according
to the French system, by regular ambulance hospitals in the
ravines leading to them. The provisional treatment should
be of the simplest kind, and chiefly directed to the prevention
of additional injury during the passage to the hospital, where
complete and accurate examination of the nature of the wound
can alone be made, and where the patient can remain at rest
after being subjected to the required treatment. The re-

moval of any missiles or foreign bodies which may be readily
obvious; the application of a piece of lint to the wound ;
the arrangement of any available support for a broken limb ;
protection against dust, cold, or other objectionable circumstances likely to occur in the transit; if "shock" exist, the
administration of a little wine, aromatic ammonia, or other
restorative, in water,—need little time in their execution,
and may prove of great service to the patient.   If hemorrhage exist from injury to a large vessel, it must of course
receive the surgeon's first and most earnest care.   He should
not trust to the pressure of a tourniquet, but secure it at
once by ligature.   Without this safeguard during the transport, and while in the hands of uneducated attendants, the
life of the wounded man might be endangered, either from
debility consequent upon gradual loss of blood or from sudden fatal hemorrhage.   It has been recommended by some
surgeons that all attendants whose duties consist in carrying
the wounded from a field of battle should be directed, when
bleeding is observed, to place a finger in the wound, and keep
it there during the transport until the aid of a surgeon is obtained.   The precise spot where compression by the finger is
wanted, and the degree of pressure necessary, will be quickly
made manifest to the sight by the effects on the flow of blood.
Such a practice seems to offer less objection than the use of
tourniquets by men whose knowledge of their proper application must be exceedingly limited.

On arrival at the hospital, where comparative leisure and
absence of exposure afford means of careful diagnosis and
definitive treatment, the following are the points to be attended to by the surgeon : firstly, examination of the wound
with a view to obtaining a correct knowledge of its nature
and extent; secondly, removal of any foreign bodies which
may have lodged ; thirdly, adjustment of lacerated structures ; and fourthly, the application of the primary dressings.

The diagnosis should be established as early as possible
after the arrival at hospital.   An examination can then be

made with more ease to the patient and more satisfactorily to the surgeon than at a later period. Not only is the sensibility of the parts adjoining the track of the ball numbed, but there is less swelling to interfere with the examination, so that the amount of disturbance effected among the several structures is more obviously apparent.

One of the earliest rules for examining a gunshot wound is to place the patient, as nearly as can be ascertained, in a position similar to that in which he was, in relation to the missile, at the time of being struck by it. In almost every instance the examination will be facilitated by attention to this precept. Occasionally it will at once indicate the probable injury to vessels or other important structures, in cases where the mutual relations of the wounds of entrance and exit, in the erect or horizontal posture of the body, would lead to no such information. Even in the direct course taken by a rifle-ball in a simple flesh wound, an erroneous opinion of the line in which the ball has moved may be formed from the first view, in consequence of the ready mobility of the several structures among themselves and their varying degrees of elasticity. Injury to nerves inducing paralysis, contusions of blood-vessels leading to secondary hemorrhage or gangrene, may thus, without sufficient circumspection, be overlooked on the first admission to hospital.

When only one opening has been made by a ball, it is to be presumed that it is lodged somewhere in the wound, and search must be made for it accordingly. But even where two openings exist, and evidence is afforded that these are the apertures of entrance and exit of one projectile, examination should still be made to detect the presence of foreign bodies. Portions of clothing, and, as has already been shown, other harder substances, are not unfrequently carried into a wound by a ball; and, though it itself may pass out, these may remain behind either from being diverted from the straight line of the wound or from becoming caught and impacted in the fibrous tissue through which the ball has passed.

The inspection of the garments worn over the part wounded
may often serve as a guide in determining whether foreign
bodies have entered or not, and, if so, their kind, and thus
save time and trouble in the examination of the wound
itself.

Of all instruments for conducting an examination of a gunshot wound, the finger of the surgeon is the most appropriate.   By its means the direction of the wound can be ascertained with least disturbance of the several structures through
which it takes its course.   If bones are fractured, the number, shape, length, position, and degree of looseness of the
fragments may be more readily observed.   In case of lodgment of foreign bodies, not only is their presence more obvious to the finger direct than through the agency of a probe
or other metallic instrument, but by its means intelligence of
their qualities is also communicated.   A piece of cloth lying
in a wound is recognized at once by a finger, while, satu-
rated with clot as it is under such circumstances, it would
probably be confounded among the other soft parts by any
other mode of examination.   The index finger naturally
occurs as the most convenient for this employment; but the
opening through the skin is sometimes too contracted to
admit its entrance, and in this case the substitution of the
little finger will usually answer all the purposes intended.
When the finger fails to reach sufficiently far, owing to the
depth of the wound, the examination is often facilitated by
pressing the soft parts from an opposite direction toward
the finger-end.

It was formerly the custom to enlarge the external orifice
of all gunshot wounds by incision, and not merely the opening, but the walls of the wound itself, as soon after the
injury as possible.   This was not done as a means of rendering the examination easier, but as a prophylactic measure.
Dilatation was also employed by tents and various other
means with a view to secure the escape of sloughs and discharges.   The opinions held by the older surgeons respect-

ing the nature of these injuries, already briefly adverted to
in the historical remarks on the subject, sufficiently explain
their object in making incisions—namely, to convert what
they regarded as a poisoned into a simple wound, and to
obviate tension, and prevent strangulation of neighboring
tissues by tumefaction on inflammation arising in its track.
Even so late as 1792, Baron Percy, in his Manuel du Chi-
rurgien d'Armée, writes: "The first indication of cure is
to change the nature of the wound as nearly as possible
into an incised one." English surgeons have, however, generally discarded the practice since the arguments used by
John Hunter against it, just about the same date as Baron
Percy wrote, excepting only in cases where it is required to
allow of the extraction of some extraneous body to secure a
wounded artery, to replace parts in their natural situation, as
in protrusion of viscera in wounds of the abdomen, or, "in
short, when anything can be done to the part wounded after
the opening is made for the present relief of the patient or
the future good arising from it." It does not often happen that
it is necessary to enlarge the openings of wounds to remove
balls, although a certain amount of constriction of the skin
may be expected from the addition of the instrument employed in the extraction ; but if much resistance is offered
to their passage out, it is better to divide the edges of the
fascia and skin to the amount of enlargement required than
to use force. In removing fragments of shells or detached
pieces of bone, the fascia and skin have almost invariably to
be divided to a considerable extent.

Where the finger is not sufficiently long to reach the bottom of the wound, even when the soft parts have been
approximated by pressure from an opposite direction, and
when the lodgment of a projectile is suspected, a long silver
probe, that admits of being bent by the hand if required, is
the best substitute. Elastic bougies or catheters are apt to
become curled among the soft parts, and do not convey to
the sense of touch the same amount of information as metal-

lic instruments do. The probe should be employed with great nicety and care, for it may inflict injury on vessels or other structures which have escaped from direct contact with the ball, but have returned, by their elasticity, to the situations from which they had been pushed or drawn aside during its passage. The above directions for examining wounds apply more particularly to such as penetrate the extremities, or extend superficially in other parts of the body; where a missile has entered any of the important cavities, search for it is not to be made, but the surgeon's attention is to be directed to matters of more vital importance to be hereafter noticed.

As soon as the presence of a ball or other foreign body is ascertained it should be removed. If it be lying within reach from the wound of entrance, it should be extracted through this opening by means of some of the various instruments devised for the purpose. In case of a leaden bullet, Coxeter's Extractor, corresponding with Baron Percy's instrument for the same purpose, and consisting of a scoop for holding and central pin for fixing the bullet, has been found a very convenient appliance, from the comparatively limited space required for its action. Instruments of two blades, or scoops, with ordinary hinge action, dilate the track of the wound injuriously before the ball can be grasped by them. The way to the removal of a bullet may often be smoothed by judiciously clearing away the fibers, among which it is lodged, during the examination, by the finger; and sometimes, by means of the finger in the wound, and external pressure of the surrounding parts, the projectile may be brought near to the aperture of entrance, so that its extraction is still further facilitated. Such foreign substances as pieces of cloth can usually be brought out by the finger alone, or by pressing them between the finger and a silver probe inserted for the purpose. Sometimes a long pair of dressing forceps, guided by the finger, is found necessary for effecting this object. Caution must be used in em-

ploying forceps, where the foreign substance is out of sight
and of such a quality that the soft tissues may be mistaken
for it.

In instances where the foreign body has not completely
penetrated, but is found lying beneath the skin away from
the wound of entrance, an incision must be made for its extraction. Before using the knife, the substance to be removed should be fixed *in situ*, by pressure on the surrounding
parts. In the instance of a round ball, the incision should
be carried beyond the length of its diameter; an addition of
half a diameter is usually sufficient to admit of the easy extraction of the ball. In removing conical balls, slugs, fragments of shells, stones, and other irregularly-shaped bodies,
the surgeon cannot be too guarded in arranging that the
fragment is drawn away with its long axis in line with the
track of the wound. By proper care in this respect, much
injury to adjoining structures may be avoided.

If balls are impacted in bone, as happens in the spongy
heads of bones, in bones of the pelvis, and occasionally,
though rarely, in other parts of long bones, they should be
removed. This can be effected by means of a steel elevator,
of convenient size; or, should this fail from the ball being
too firmly impacted, a thin layer of the bone on one side of
the ball may be gouged away, so that a better purchase may
be obtained for the elevator, in effecting its removal. The
fact is now fully established that, although in a few isolated
cases balls remain lodged in bones without sensible inconvenience, in the majority the lodgment leads to such disease
of the bony structure as often to entail troublesome abscesses,
and in some instances eventually to necessitate amputation.
The lodgment of balls will not often occur without extensive
fracture in warfare where rifled arms of such force as the
Minié or Enfield are the chief weapons employed, but will
not unfrequently be met with in such campaigns as have
lately happened in India.

6

Should there be reason for concluding that a ball or other foreign body has lodged, but after manual examination, and observation as well by varied posture of the part of the body supposed to be implicated as by indications derived from the patient's sensations, effects of pressure or injury to nerves, and all other circumstances which may lead to information, should the site of the lodgment not be ascertained, the search should not be persevered in to the distress of the patient. Neither, although the site of lodgment be ascertained, if extensive incisions are required, or if there is danger of wounding important organs, should the attempts at extraction be continued. Either during the process of suppuration, by some accidental muscular contraction, or by gradual approach toward the surface, its escape may be eventually effected; or, if of a favorable form, and if not in contact with nerve, bone, or other important organ, it may become encysted, and remain without causing pain or mischief. When John Hunter wrote on gunshot wounds, he remarks, The practice of searching after a ball, broken bones, or any other extraneous bodies, had been in a great measure given up, from experience of the little harm caused by them when at rest, and not in a vital part; and he himself advises, even when a ball can be felt beneath skin that is sound, that it should be let alone, chiefly on the ground that two wounds are more objectionable than one, and that the extent of inflamed surface is proportionably increased by incision. More extensive experience has, however, shown that not only is the risk of subsequent ill results greater in those cases where foreign bodies remain lodged than when they have been cut out, but also that the advantages of a second opening for the escape of the necessary sloughs and discharges greatly preponderate over the disadvantages connected with it, as regards the additional extent of injured surface. The advantage also of the satisfaction to the mind of a patient from whom a ball has been removed must not be overlooked; for men suffering from gunshot wounds are invariably rendered uneasy by a vague apprehension of danger, for some time after the injury, if the missile has remained undiscovered.

When a gunshot wound has been accompanied with much laceration and disturbance of the parts involved in the injury, it is necessary, after the removal of all foreign substances that can be detected, to readjust and secure the disjointed structures as nearly as possible in their normal relations to each other. The simplest means—strips of adhesive plaster, light pledgets of moist lint, a linen roller, favorable position of the limb or part of the body wounded—should be adopted for this purpose. Pressure, weight, and warmth should be avoided as much as possible in these applications, consistent with the end in view. It must not be forgotten, in thus bringing the parts together, that the purpose is not to obtain union by adhesion, which cannot be looked for, but simply to prevent avoidable irritation and malposition of parts, during the subsequent stages of cure by granulation and cicatrization. In all gunshot wounds, much discomfort to the patient is prevented by carefully sponging away all blood and clot from the surface adjoining the wound, and by adopting measures to prevent its spreading again in consequence of oozing. This can be readily done with the aid of a little warm water, and arrangement when the wound is first dressed, but can only be accomplished with considerable inconvenience after the thin clots have become hard and firmly adherent to the skin.

When the parts of a lacerated gunshot wound have been brought into apposition, as in simple penetrating wounds, the only dressing necessary is moistened lint. It should be kept moist either by the renewed application of water dropped upon it, or by preventing evaporation by covering it with oiled silk The sensations of the patient may be consulted in the selection of either of these, and climate and temperature will be often found to determine the choice. In hot climates cold applications are the more grateful, and by checking the amount of inflammatory action and circumscribing its extent are usually the more advantageous.   M.

Velpeau and other French surgeons have strongly recommended the use of linseed-meal poultices, above all wet linen applications. Charpie is still extensively employed in French military hospitals.* M. Baudens and Dr. Stromeyer have strongly recommended the topical application of ice placed in bladders ; others, the continued irrigation of the wound with tepid water. The means of applying such remedies are rarely available in the military hospitals where gunshot wounds are ordinarily treated in their early stages. When much local inflammation has set in, and when there is much constitutional fever even without unusual local irritation, the non-evaporating or warm applications will be found to be the most advantageous.

When suppurative action has been fully established, the

---

* M. Scrive gives the following as the weight of the linen dressings consumed by the wounded of the French army in the campaign in the Crimea :—

		English weight. tons. cwt. qr. lb.			
Linen cloth............101,779 kilogrammes	=	100	2	1	23
Rolled bandages..... 46,446 "	=	45	13	2	14
Charpie .... 47,776 "	=	46	19	3	4

And estimates the following as the proportion consumed by each of the wounded :—

		English weight avoirdupois. lb. oz. dr. gr.			
Linen cloth...............2 kil. 482 grammes	=	5	7	0	10
Rolled bandages.........0 " 891 "	=	1	15	7	13
Charpie ....1 " 181 "	=	2	9	11	0
Total.................4 " 554 "	=	10	0	2	23

In an Army Medical Department Circular, dated 27th May, 1855, it was announced that the Secretary of State for War had decided the following "Field Dressing" should form part of every British soldier's kit on active service, so as to be available at all times and in all places as a first dressing for wounds :—

Bandage of fine calico, 4 yds. long, 3 in. wide.
Fine lint, 3 in. wide, 12 in. long.
Folded flat and fastened by 4 pins.

surgeon must be guided by the general rules applicable to all other such cases. Care must be taken to prevent the accumulation of pus, lest it burrow, and sinuses become established—not an unfrequent result of want of sufficient caution in this regard. If much tumefaction of muscular tissues beneath fasciæ occurs, or abscesses form in them, free incisions should be at once made for their relief. In wounds where the communication between the apertures of entrance and exit is tolerably direct, occasional syringing with tepid water may be useful, by removing discharges and any fibers of cloth which may be lying in the course of the wound. Weak astringent solutions are occasionally employed in a similar way, with a view to improving the tone of the exhalents and exciting a more vigorous action in the process of granulation. The strictest attention to cleanliness and the complete removal of all foul dressings are essentially necessary, not merely for the comfort of the patient, but to prevent the accumulation of noxious effluvia, and also to obviate the access of flies to the wounds. In tropical climates, and in field-hospitals in mild weather, where many wounded are congregated, flies propagate with wonderful rapidity, and the utmost care is necessary to prevent the deposit of ova and generation of larvæ in the openings of gunshot wounds, especially while sloughs are in process of separation. Cloths dipped in weak solutions of creasote or disinfecting fluids, laid over the wound, are found necessary for this purpose when the insects abound in great numbers.

The constitutional treatment in an ordinary gunshot wound, uncomplicated with injury to bone or structures of first importance, should be very simple. The avoidance of all irregularity in habits tending to excite febrile symptoms or to aggravate local inflammation, attention to the due performance of the excretory functions, and support of the general strength, are chiefly to be considered. Bleeding, with a view to prevent the access of inflammation in such cases, is now never practiced, as formerly, by English surgeons.

6*

The diet should be nutritious, but not stimulating. A pure fresh atmosphere is a very important ingredient in the means of recovery. If from previous habits of the patient, or from circumstances to which he is unavoidably exposed, the local inflammation has become aggravated,—indicated by pain, increased swelling, and redness about the wound,—topical depletion by leeches or cupping, bleeding from the arm, saline and antimonial medicines, and strict rest in the recumbent position, must be had recourse to, the extent being regulated by the circumstances of each case. In instances such as these, when the inflammation has become diffused, the purulent secretion is not confined to the track of the wound, but is liable to extend among the areolar connections of the muscles; and if the cure be protracted, attention will be necessary to prevent the formation of sinuses. If stiffness or contractions result, attempts must be made to counteract them by passive motion and friction, with appropriate liniments; if a tendency to edema and debility remain in a limb after the wound is healed, the cold-water douche will be found to be one of the most efficient topical remedies. In French practice, the administration of a chalybeate tincture,* as a tonic, or diluted as an injection, in wounds threatening to assume an unhealthy character, is very highly praised. It is stated that under the conjoined employment of this remedy internally and externally, in wounds of a pallid, unhealthy aspect, accompanied by nervous irritability and symptoms of approaching pyemia, the granulations have resumed a red and healthy appearance, and the general state of health become rapidly favorable.

Progress of cure.—Simple flesh wounds from gunshot usually heal in five or six weeks. In the course of the first day the part wounded becomes stiff, slightly swelled, tender, a slight inflammatory blush surrounds the apertures through

---

* Perchlorure de fer, 30 drops, two or three times daily as a tonic, and diluted with six parts of water as an injection.

which the missile has passed, and a slight serous exudation escapes from them. Suppuration commences on the third or fourth day, and in about ten days or a fortnight the sloughs are thrown off. Granulation now progresses, more or less quickly according to the health and vigor of the patient's constitution. The opening of exit is usually the first closed. When the wound is complicated with unfavorable circumstances, whether inducing in the patient a condition of asthenia or leading to excess of inflammatory action, the progress of the cure may be extended over as many months as, under favorable circumstances, weeks are occupied in the process.

GUNSHOT WOUNDS IN SPECIAL REGIONS OF THE BODY.

The circumstances connected with wounds in particular situations of the body, or in particular organs, are in many respects common to injuries from other causes than gunshot; and in the following remarks the attention is chiefly drawn only to those leading peculiarities which constantly demand the consideration of the army surgeon, and which spring either from the nature of gun projectiles, or the circumstances under which this branch of military practice has for the most part to be pursued.

GUNSHOT WOUNDS OF THE HEAD.

No injuries met with in war require more earnest observation and caution in their treatment than wounds of the head. The vital importance of the brain; the varied symptoms which accompany the injuries to which this organ may be subjected, directly or indirectly; the difficulty in tracing out their exact causes; the many complications which may arise in consequence of them; the sudden changes in condition which not unfrequently occur without any previous warning,—all these circumstances will keep a prudent surgeon who has charge of such wounds continually on the alert. Injuries of this class, the most slight in appearance at their onset, not unfrequently prove most grave as they proceed, from encephalitis and its consequences, or from plugging of the sinuses by coagula, leading to coma, paralysis, or pyemia; and the converse sometimes holds good with injuries presenting at first the most threatening aspects, where care is taken to avert these serious results.   Much will depend on the part of the head struck, both as regards the thicker and stronger processes or portions of the skull, and the situation of the sinuses and parts of the cerebrum within; on the force and shape of the projectile; the angle at which it strikes; the age and condition of the patient; and other matters already referred to in the general remarks on gunshot wounds.   Mr. Guthrie has laid down as a rule that injuries of the head, of apparently equal extent, are more dangerous on the forehead than on the side or middle portion, and still more so than those on the back part; and that a fracture of the vertex is infinitely less important than one at the base of the cranium.   When the injuries are caused by rifle-balls, however, these considerations are rarely of much avail, for the power of injury is such that it can scarcely ever be confined to the immediate neighborhood of the part directly struck.

Wounds of the head may be divided, for convenience of description, into wounds of the scalp and pericranium, without fracture of bone ; similar wounds complicated with fracture of the outer or of both tables, without pressure on the encephalon; wounds with fracture and depression; and lastly, wounds in which the encephalon itself has been penetrated.   Severe contusion of the bones of the cranium, followed by necrosis, and even fracture, with or without depression, may occur without an open wound of the superficial investments.   The case of an officer is mentioned in Dr.

Macleod's Notes of the Crimean War, who was thus killed by a round shot. The scalp was not cut, almost uninjured, but the skull was most extensively comminuted.

Wounds of the scalp and pericranium.—These wounds are usually inflicted by projectiles which are brought into contact at a very acute angle, so that little direct injury to the brain or its membranes is inflicted, and the surgeon's attention need only be directed to the same considerations as must occur in any contused wounds of the scalp from other causes than gunshot. But even in these accidents, though appearing to be simple flesh wounds, serious cerebral concussion and other lesions are occasionally met with. The usual stupor and other signs of concussion may be very evanescent, or may last for several days, disappearing gradually and wholly, or entailing subsequent evils at more or less remote periods. It must not be forgotten that when the pericranium is removed by a musket-ball, however superficial the injury may seem, there is always a certain degree of injury and bruising to the bone from which it is torn, and necessary laceration of the vessels which inosculate with the nutritive capillaries of the diploë, and through them of the vessels of the meninges with which they are connected. The injury to this vascular system almost invariably leads to necrosis of the portion of the skull from which the coverings are carried away; and sometimes, even when the pericranium is not torn off, sufficient injury is inflicted to lead to a like result. The death of bone is generally limited to a thin layer of the outer table, which in due time exfoliates. The injury to the vessels ramifying between the inner surface of the cranium and dura mater may lead to serious results. There may be rupture of a sinus, leading to compression, or fatal results may ensue from inflammation and suppuration. The case of a young soldier in whom the longitudinal sinus was thus ruptured occurred to the writer. In this instance a rifle-ball had divided the scalp and pericranium about four inches in length obliquely across the skull, just anterior to the angle of the lambdoidal suture, the posterior end of the sagittal suture being exposed midway in the line of the wound. The patient vomited at the instant of the blow, and symptoms of compression, mixed with some of concussion, soon followed. He died eleven hours after the injury. At a post-mortem examination, the superior longitudinal sinus was found to be ruptured, and about four ounces of coagulated blood were lying on the brain. Two darkly-congested spots were observed in the cerebrum, one on each hemisphere, corresponding with the line of direction in which the ball had passed, and these, when cut into, presented the usual characters of ecchymoses. There was no fracture of bone. The case may be found detailed at some length in the *Lancet*, vol. i., 1855. When inflammation follows the passage of a ball, whether terminating in resolution or leading to abscess, the symptoms and treatment required will be the same as in similar affections from other causes. In like manner, the occurrence of erysipelas, or other complications to which these wounds of the scalp are liable, will be found treated elsewhere. (See INJURIES OF THE HEAD.)

The treatment of an ordinary gunshot wound of the scalp should be very simple. Cleansing the surface of the wound, removing the hair from its neighborhood for the easier application of dressings, lint moistened with clean water, very spare diet, and careful regulation of the excretions are the only requirements in most cases. The patient must be closely watched, so that measures may be taken to counteract inflammatory symptoms in their earliest stages. Even after one of these wounds has healed, and the patient to all appearance has quite recovered, it is necessary to enjoin continued abstinence from excesses of all kinds. Instances are frequently quoted where intoxication, a long time after the date of injury, has induced symptoms of apoplexy and death. In the Surgical History of the Crimean Campaign, the case of a soldier of the 31st Regiment, thirty-eight years old, who received a contused wound at the back of the head from a piece of shell, without section of the scalp and without lesion of the bone, is related. In this instance a small abscess formed under the scalp, and was evacuated. After the wound was healed the man suffered from constant headaches, and was invalided to England. Soon after landing he drank freely, coma followed, and he died shortly afterward. The post-mortem examination showed traces of inflammatory action in the dura mater, and "just anterior and superior to the corpora quadrigemina was a tumor the size of a walnut, composed of organized fibrin and some clotted blood."

Wounds complicated with fracture, but without depression on the cerebrum.—These are very uncertain in their effects, and often apt to mislead the surgeon, from the absence of urgent symptoms in their early stages. The occurrence of fracture is, however, sufficient to show the force with which the projectile has struck the head, and to indicate the mischief which the brain and its immediate coverings have not improbably sustained.

In these injuries there may be a simple furrowing of the outer table, without injury to the inner; or there may be fissure extending to a greater or less degree of length, or radiating in several lines ; or both tables may be comminuted in the direction the ball has traversed in such small portions that they lie loosely on the dura mater without much alteration in the general outline of the cranial curve. The chief and only means, in many cases, of concluding that no depression upon the cerebrum has taken place is the absence of the usual symptoms of compression ; for it is well known that simple observation of the injury to the outer table, whether by sight or touch, will by no means necessarily lead to a knowledge of the amount of injury or change of position in the inner table.

When simple removal of a portion of the outer surface of the skull has been caused by the passage of the ball or other missile, the wound will sometimes heal, under judicious treatment, without any untoward symptom. A layer of the exposed surface of bone will probably exfoliate, and the wound granulate and become closed without further trouble. But such injuries, for reasons before named, are very likely to be followed by inflammation, and not improbably abscess, between the internal table and dura mater; and further, as a consequence of the vascular supply being stopped, and perhaps also partly from the effects of the original contusion by necrosis of the inner table itself. Care must be taken not to mistake one of these injuries for a depressed fracture, as is not unlikely to happen when the excavation effected by the projectile is rather deep and the edges of the bone bordering the excavation are sharp.

Fissured fractures, when the fissure extends through the skull, usually result from injuries by shell. The passage of a ball may fracture and very slightly depress a portion of the outer table of the cranium, and then the line of fracture will very closely simulate fissured fracture extending through both tables, and the diagnosis between them be excessively doubtful. When fissured fracture exists, the distance to which it may be prolonged is often quite unindicated by symptoms, and its extent is very uncertain. Fissures often extend to long distances. They may occur at a part remote from the spot directly injured. In the case of a lieutenant of the 11th Hussars, who was apparently slightly wounded at Balaklava in the middle of the forehead by a piece of shell, a fissured fracture was found, after death, across the base of the skull, quite unconnected with the primary wound, and seemingly from *contre-coup*. Death resulted from inflammation and suppuration set up near this indirectly-injured part. Fissured fracture of the inner table may also occur from the action of a ball without external evidence of the fracture. Such a case occurred in the 55th Regiment, in the Crimea. The soldier had a wound of the scalp along the upper edge of the right parietal bone. The ball in passing had denuded the bone; but there was no depression. The

man walked to camp from the trenches without assistance,
and there were no cerebral symptoms on his arrival at hospital; but five days afterward there was general edema of
the scalp and right side of face, the wound became unhealthy,
and slight paralysis appeared on the left side. The next
day hemiphlegia was more marked, convulsion and coma
followed, and he died on the thirteenth day after the injury.
Pressure from a large clot of coagulum and extensive inflammatory action were the immediate causes of death; but
a fissure, confined to the inner table, running in line with the
course of the ball, was also discovered. A preparation of
the calvarium in this case was presented by Dr. Cowan, 55th
Regiment, to the museum at Fort Pitt.

The cases where comminution has resulted from the track
of a ball across the skull generally present less unfavorable
results than those where a single fissured fracture, extending
through both tables, exists. The small, loose fragments can
be removed; and if the dura mater be intact, the case, with
proper care to prevent inflammatory action, may not improbably be attended with a favorable recovery.

**Wounds complicated with fracture and depression on
the cerebrum.**—Such wounds are most serious, and the
prognosis must be very unfavorable. They must not be
judged of by comparison with cases of fracture with depression caused by such injuries as are usually met with in civil
practice. The severe concussion of the whole osseous sphere
by the stroke of the projectile, the bruising and injury to the
bony texture immediately surrounding the spot against which
it has directly impinged, as well as the contusion of the external soft parts, so that the wound cannot close by the ad-
sive process, constitute very important differences between
gunshot injuries on the one side, and others caused by instruments impelled solely by mnscular force on the other. So,
also, the injury to the brain within, and its investments, is
proportionably greater in such injuries from gunshot. The
experience of the Crimean campaign shows that, when these

7

injuries occurred in a severe form, they invariably proved fatal. Of seventy-six cases treated, where depression only, without penetration or perforation, existed, fifty-five proved fatal, twelve were invalided, and nine only were discharged to duty. In the twenty-one survivors, the amount of depression is stated in the history of the campaign to have been slight, though unmistakable, and all except one recovered without any bad symptom. Of eighty-six other cases where perforation or penetration of the cranium occurred, all died.

With penetration of the cerebrum.—It is obvious that, where a projectile has power not only to fracture, but also to penetrate the cranium, it will rarely be arrested in its progress near the wound of entrance. Either splinters of bone, or the ball, or a portion of it will be carried through the membranes into the cerebral mass. Sometimes a ball, if not making its exit by a second opening in the cranium, will lodge at the point of the cerebral substance opposite to that of its place of entrance ; but the course a projectile may follow within the cranium is very uncertain.

Instances have occurred where balls have lodged in the cerebrum without giving rise to serious symptoms of danger for a long time. Such cases might lead to throwing surgeons off their guard in making a prognosis, from supposition that the ball by some accident had not lodged. The case of a soldier wounded by a ball in the posterior part of the side of the head is mentioned by Mr. Guthrie. The wound healed, and the man returned to duty ; a year afterward he got drunk, and died suddenly. The ball was found in a sac lying in the corpus callosum. Another soldier wounded at Waterloo had a similar recovery, and also died after intoxication. The ball was found deeply lodged in a cyst in the posterior part of the brain. An artillery soldier was wounded, in the Crimea, by a rifle-ball, which entered near the inner angle of the left superciliary ridge. The wound progressed without a bad symptom until a month afterward, when coma came on, and death shortly followed. The ball was found in a sac, in which pus also was contained, at the base of the left anterior lobe of the brain.

Treatment.—The treatment of the various kinds of fractures from gunshot, and their complications, may be considered together. Formerly, a gunshot wound of the head was supposed to be in itself a sufficient indication for the use of the trephine; indeed, even where no fracture was caused, an opening was recommended by comparatively recent surgeons to be made in the cranium, to meet symptoms which might be expected to result. Modern surgeons, however, generally have made use of the trephine only when there was reason for concluding that depressed bone was leading to *permanent* interruption of cerebral function, or that an abscess had formed within reach, and was capable of evacuation. Preventive trephining has been proved to be useless, as well as dangerous, and is no longer an admissible operation. The tendency of the most recent experience has been to limit the practice of trephining to the narrowest sphere; and when the very great difficulty of making accurate diagnosis in these cases is considered,—whether as to the distinguishing signs of compression; the precise seat of its cause, if the compression exist; the space over which this cause, when ascertained, may extend; its persistent or temporary character; its complications; and certain dangers connected with the operation itself,—no wonder need be excited that this tendency should exist. Besides, the numerous cases which have now been noted where bone has evidently been depressed, but the brain has accommodated itself to the pressure without serious disability being caused, or where compression from effusion has been removed by absorption under proper constitutional treatment, are further causes of hesitation in respect to trephining. In the Surgical Report of the Crimean Campaign, it is stated that the trephine was only successfully applied in four cases (and none of these were from rifle-balls) during the whole war; and

that in these instances the patients were subsequently subject to occasional headache and vertigo ; and in the French
report, by Dr. Scrive, it is stated that trephining was for
the most part fatal in its results in the French army. In
siege operations, the experience as regards wounds of the
head is always very extensive, the lower parts of the body
being so much more protected in the trenches. According
to Dr. Scrive's returns, one of every three men killed in the
trenches before Sebastopol, and one in every 3·4 wounded,
was injured in this region. In the English returns, wounds
of the head and face in the men are shown as 19·3 per cent. ;
in the officers, as 15 per cent. ; but this is of the total
wounded in the field as well as in the trenches. There was,
therefore, as extensive a range for observation of the effects
of trephining in the siege of Sebastopol as is likely to happen in any war. Dr. Stromeyer, who in the early part of
his professional career resorted to trephining in complicated
fractures of the skull, records, in his Principles of Military
Surgery, that he has abandoned the practice. After the
battle of Kolding, in Sleswick, in 1849, there were eight
gunshot fractures of the skull, with depression, and more or
less cerebral symptoms. In all these, with one exception,
the detachment of the fractures was left to nature, and all
recovered. One patient, from whom some fragments were
removed on the seventh day, was placed in considerable
danger by the treatment, and Dr. Stromeyer resolved never
to adopt it again. In 1850, in Sleswick, two young surgeons came under Dr. Stromeyer's care with gunshot wounds
of the head, accompanied by deep depression ; they were
both treated without trephining, and both recovered.
Throughout the three campaigns of the Sleswick-Holstein
war, there was only one case of trephining which gave a
favorable result. Military experience makes it difficult to
understand the frequent and successful performance of tre-
panning by the older surgeons for such slight causes as they
performed it, excepting that the patients labored under

little else than the effects of the operation itself, while very fatal mischief has existed in addition in those instances in which the operation has been resorted to for accidents from gunshot. A circumstance quoted by Sir G. Ballinghall particularly illustrates the favorable results of abstaining from trephining in some cases. After the battle of Talavera, a hospital which had been established in the town had to be suddenly abandoned, and an order was given for all the wounded who could march to leave it. There was no time for selection, and among those who marched were twelve or fourteen men with wounds of the head, in which the cranium was implicated, four or five having both tables fractured, and two having the globe of one eye destroyed along with fracture of the os frontis. All these men recovered, though they were sixteen days on the march, harassed and exposed to a burning sun, and had no other application than water-dressing. Of eight cases of contusion or fracture of the cranium, with displacement of both tables, recorded by Dr. Williamson, among men who were sent from India to Chatham, during the late mutiny, none had been trephined. In all these there was a depressed cicatrix, the wound having contracted and become closed by a strong fibrous investment. In one case—a wound by a musket-ball, in the center of the forehead—the ball was supposed to be still lodged within the skull. In the Fort Pitt museum are several preparations, showing depressed fracture of the inner table of the skull from gunshot, taken from patients who had recovered without trephining, and died years afterward from other causes. The edges of the depressed portions of bone had become smooth, and united by new osseous matter, and the cerebrum must have accommodated itself to the new form of the inner cranial surface. Two or three instances are known in which the course of a ball has been traced from the sight of entrance across the brain, and trephining resorted to for its extraction, with success; but there are also many others in which the mere operation of the extraction of a foreign body has apparently led to the immediate occurrence of fatal results. Moreover, splinters of bone are not unfrequently carried into the brain by balls, and these may elude observation; or the ball itself may be divided and enter the brain in different directions, as was observed in the Crimea; when the operation of trephining can only be an additional complication to the original injury, without any probable advantage. Where irregular edges, points, or pieces of bone are forced down and penetrate—not merely press upon—the cerebral substance, or where abscess manifestly exists in any known site, or a foreign substance has lodged near the surface, and relief cannot be afforded by the wound, trephining may be resorted to for the purpose; but the application of the operation, even in these cases, will be very much limited if certainty of diagnosis be insisted upon.   In all other cases, it seems now generally admitted that much harm will be avoided, and benefit more probably effected, by employing long-continued constitutional treatment, viz., all the means necessary for controlling and preventing the diffusion of inflammation over the surface of the brain and its membranes,—the most careful regimen, very spare diet, strict rest, calomel and antimonials, occasional purgatives, cold application locally, so applied as to exclude the air from the wound, and free depletion by venesection, in case of inflammatory symptoms arising. Similar remarks will apply in case of lodgment of a projectile within the brain; if the site of its lodgment is obvious, it should be removed with as little disturbance as possible, but trephining for its extraction on simple inference is unwarrantable.

GUNSHOT WOUNDS OF THE SPINE.

Gunshot wounds of the spine are closely associated with similar injuries of the head. In both classes corresponding considerations must be entertained by the surgeon in reference to the important nerve-structures, with their membranes, which are likely to be involved in the injury to their osseous envelope; in both, the effects of concussion, compression, laceration of substance, or subsequent inflammatory action, chiefly attract attention. In the Surgical History of the Crimean Campaign, twenty-seven cases are noted in which vertebræ were fractured, eight being without apparent lesion of the spinal cord, and nineteen with evident lesion. Of these, twenty-five died; and two, in which the fractures were confined to the processes of the vertebræ, survived to be invalided. The gunshot wounds affecting the spinal column have not been separated from injuries in other regions in the French returns. Six men only wounded in the spine, during the late mutiny in India, arrived in Chatham. In all, they were the results of musket-balls. Two were wounds of the sacrum; in the remainder, the portions of the vertebræ fractured were the spinous processes. Concussion of the spinal column, leading to paralysis more or less persistent, is usually occasioned by fragments of shell, or stones from parapets; and in these cases the accidents are mostly accompanied by extensive lesions of the neighboring structures. In one fatal case in the Crimea, the ball passed through the spine rather below the first dorsal vertebra, leading to complete loss of sensation and voluntary motion below the seat of injury, and death on the sixteenth day afterward; in another, a rifle-bullet entered the right side of the second lumbar vertebra, traversed the spinal canal at that part, and lodged in the body of the bone. In this latter case, violent pain was complained of in the lower extremities, shooting along the groins. The patient was paraplegic, and death ensued thirty-three hours after admission. In another fatal case, a rifle-bullet passed through the right cheek, and lodged near the base of the skull. There was no paralysis, but delirium and coma supervened, and the patient died five days after receiving the wound. The bullet was found after death, lying just below the basilar process, and a large piece of the atlas was broken off and almost detached. The spinal cord did not appear to have been primarily injured, but acute inflammation had been set up, and had extended to the membranes of the brain. There is a preparation in the museum at Fort Pitt which shows fracture both of the atlas and axis, without lodgment of the ball. The patient survived thirty days. It is curious that, in a case under the care of the writer, before referred to, where a rifle-ball passed through the right loin, entered the spinal canal between the third and fourth lumbar vertebræ, breaking the laminæ, passed upward within the column, between it and the cord, and made its exit through the left intervertebral foramen between the second and third vertebræ, as shown after death, no paralysis occurred at the time of the injury, nor subsequently, nor was any evidence afforded post mortem of thecal inflammation having been excited. (See Guy's Reports, vol. v., 1859.)

In injuries of the vertebral column and spinal cord occurring in military practice, the mischief is usually so complicated and extensive, and the medulla itself so bruised, that the cases must be very rare indeed in which the operation of trephining, if justifiable in any case, can offer the slightest prospect of benefit. M. Baudens extracted, with an elevator supplied with a canula, a ball which had lodged in the eleventh dorsal vertebra and was causing compression with complete paraplegia. The paralysis disappeared immediately after the extraction of the bullet; but tetanus came on four days afterward, and proved speedily fatal. Balls have been known to pass through the bodies of vertebræ, and apparent cure follow; but as such patients in military practice

are usually invalided out of the service as soon as they are
fit to leave hospital, no opportunity is afforded of observing
the consequences which ulteriorly ensue.

GUNSHOT WOUNDS OF THE FACE.

Wounds of the face from musket-shot, grape, and small
fragments of shell are usually more distressing from the de-
formity they occasion than dangerous to life. The absence
of vital organs, the natural divisions among the bones, and
their comparatively soft structure, rendering them less liable
to extensive splitting; the copious vascular reticulation and
supply rendering necrosis so much less likely and repair so
much easier than in other bones; the limited amount of space
occupied by the osseous structure between their respective
periosteal investments, and the opportunities from the number of cavities and passages connected with this region for
the escape of discharges, lead to this result. On the other
hand, the vascularity of this region leads to danger both of
primary and especially secondary hemorrhage—a circumstance which, in all deep wounds of this region, must be
looked for as a not improbable complication. The other
complications of these gunshot wounds are lesions of the
organs of special sense, injury to the base of the skull, paralysis from injury to nerves, wounds of glands, their ducts,
and of the lachrymal apparatus; but it is scarcely necessary
to do more than allude to them, as the considerations connected with their treatment will be found elsewhere.

Wounds from cannon-shot occasionally illustrate what terrible injuries may be borne in this region without life being
at once extinguished. They are the more distressing because
the patient lives conscious of his sufferings without possibil-
ity of surgical alleviation. The case of an officer of Zouaves,
wounded in the Crimea, is recorded, who had his whole face
and lower jaw carried away by a ball, the eyes and tongue
included, so that there remained only the cranium, supported

by the spine and neck. This unfortunate being lived twenty
hours after the injury, breathing by the laryngeal opening at
the pharynx, while his gestures left no doubt that he was conscious of his condition. Mr. Guthrie has recorded a similar
case which occurred in an officer during the assault of Bad-
ajos. This patient suffered distressingly from want of water
to moisten his throat, but could not swallow when some was
brought. One eye was left hanging in the orbit, the floor
of which was destroyed, and this enabled him to write thanks
for attention paid him. He did not die till the second night
after the injury.

In the treatment of gunshot wounds of the face where the
bones are splintered and torn, the surgeon should always re-
tain and replace as many of the broken portions as possible.
It is often surprising how small connections with neighboring soft parts will suffice to maintain vitality and lead to restored union in this region. A case which occurred to the
writer in August, 1855, in a private of the 19th Regiment,
is detailed in the *Lancet*, p. 436, of that year. The wound
was caused by a fragment of shell. The right half of the
arch of the palate was jammed in and fixed at right angles
to the other half, and the upper maxillary bone was so comminuted that it was scarcely possible to note the directions of
the lines of fracture. The lower maxilla was broken in three
places, and there was extensive laceration of the soft parts.
Great difficulty was met with at first in unlocking the parts
of the palate which had been driven into each other, and,
when they were separated, the right half hung down loosely
in the mouth; yet favorable union was obtained between all
these fractures, the broken portions being adjusted so that
the man recovered with both the upper and lower maxillæ
consolidated in their normal relations to each other. No teeth
had been driven out of their sockets, and they were very
useful as points of support in the steps taken to procure
coaptation of the disunited fragments. In the *Lancet* of
February 24th, 1855, may be found the description of a se-

ries of wounds of the face, from the Crimea, which were examined by Mr. Samuel Solly, and described by him, some of them illustrating how wonderfully the larger arteries often escape in these injuries. In one, loss of the sense of taste on one side of the tongue had resulted; in two, there was partial paralysis of the portio dura; in another, impaired action of the jaw. In one, where a ball entered at the junction of the malar bone and os frontis on the left side, and descended and escaped at the posterior border of the sterno-mastoid muscle, the sight of the left eye was destroyed, and that of the right weakened; and constant headache, dullness of intellect, and incapacity for mental application remained. The injury had originally been followed by symptoms of cerebral concussion. In another case, the man came home with an iron shot firmly wedged and lodged in the center of the vomer. When extracted, at Chatham, by Staff-Surgeon Parry, it was found to weigh nearly four ounces. The returns of the Crimean campaign, from the 1st of April, 1855, to the end of the war, show 533 wounds of the face, of which number 445 returned to duty, 74 were invalided, and 14 died. Bones were penetrated in 107 of these cases, one eye was injured in 42, and both eyes in 2 cases. Mr. Guthrie has recorded that he several times saw both eyes destroyed by one ball, without much other mischief, and one, and even both, rendered amaurotic by balls which had passed behind the eyes. Of 21 cases of wounds of the face, with injuries to bones, returned to England from the late Indian mutiny, and recorded by Dr. Williamson, 11 had lost the sight of one eye, and 1 of both eyes; 6 cases were complicated with fracture of the lower jaw, and in 3 of these the fracture remained ununited.

GUNSHOT WOUNDS OF THE CHEST.

These always form a large proportion of the injuries from warfare, both in the open field and more especially in sieges, where the upper part of the body is chiefly exposed. Dr. Scrive's returns show that the proportion of chest to other wounds was 1 in 12 in the trenches, and 1 in 20 in ordinary engagements. In the British forces they are returned as 1 in 10 among the officers during the whole war, and nearly 1 in 17 among the men, from 1st April, 1855, to the end of the war. The ample space of this region, and the exposed surface it offers as a target toward the enemy, would lead to an anticipation of such results. The serious complications which ensue when the cavity of the chest is penetrated, and the dangerous consequences of wounds of its viscera, cause the proportionate mortality to be very great. The British returns show that among the officers treated for these wounds $31\frac{1}{2}$ per cent. and among the men $28\frac{1}{10}$ per cent. died. Out of 603 wounded men who returned to England from the late Indian mutiny, the number who had received wounds of the chest was only 19. In many instances men thus wounded do not live long enough to come under treatment, but die on the field of action from penetration of the heart, hemorrhage, suffocation, or shock ; and the proportion of chest wounds returned as "killed in action," or as "died under treatment," will constantly vary according to circumstances connected with the nature of the military operations, and the opportunities of early removal from the field to hospital.

Gunshot wounds of the chest may conveniently be divided for study into two classes, viz., *non-penetrating* and *penetrating*. NON-PENETRATING wounds become subdivided into simple contused wounds of the soft parietes ; contused and lacerated wounds ; the same accompanied with injury to bones or cartilage ; and, lastly, those complicated with lesion of some of the contents of the chest, the pleura remaining

unopened, or, if opened, without a superficial wound. PENE-
TRATING wounds may exist without wound, or with wounds
of one or more of the viscera of this cavity. Among the
more serious complications with which the latter may be
accompanied is the lodgment of the projectile or other foreign bodies, as of fragments of bone, within the chest. As
wounds of the heart and great vessels are almost invariably
at once fatal, and as the organs of respiration occupy the
greater part of the cavity of this region, it is in reference to
the latter that the treatment of chest wounds is chiefly con-
cerned.

Non-penetrating wounds.—Of the simpler wounds in
which the soft parietes only are involved little need be observed, excepting that the healing process is often prolonged
by the natural movements of the ribs to which the wounded
structures are attached, especially when the ball has taken a
circuitous course beneath the skin, and that the surgeon must
be on his guard to watch for pleuritis arising as an occasional
consequence of these injuries. In two deaths recorded in
the Director-General's History of the Crimean War, under
simple flesh wounds, without fracture or pleural opening,
from bullets, the fatal termination arose from pleuro-pneu-
monia. When the force has been great, as when fragments
of shell or rifle-balls strike at full speed against a man's
breast-plate, not only may troublesome superficial abscesses
and sinuses follow, but the lungs may have been compressed
and ecchymosed at the time of the injury, and hemoptysis
be one of the symptoms presented.

When the projectile has been of large size, although no
opening of the parietes or fracture exists, death sometimes
ensues by suffocation as the direct result of pulmonary en-
gorgement. The danger of pleuritis or pneumonia will be
greater when the injury has been so severe as to cause
division of bone or cartilage, and the subsequent suppuration and process of exfoliation will not unfrequently prove
very tedious and troublesome. Although the pleura has

not been opened, the lung may be lacerated either by the force of contusion or, as in a case recorded by Dr. Macleod, by the edges of the fractured ribs, which may afterward return to their normal relative positions, so as to leave no indication during life of the means by which the lung had been wounded. Such an injury would be rendered much more probable by the existence of old adhesions, connecting the pulmonary and costal pleuræ opposite to the site of injury.

Notwithstanding a projectile has not penetrated the parietes of the chest, a pleural cavity may be opened, as in injuries from other causes, and the lung wounded by the sharp edges of fractured ribs. This will be indicated by emphysema, pneumothorax, hemoptysis, probably signs of internal hemorrhage, and inflammation. Such wounds will generally be the result of injuries from fragments of shell.

Penetrating wounds.—These wounds, especially when the lung is perforated or the projectile lodges, are necessarily exceedingly dangerous. Fatal consequences are to be feared, either from hemorrhage, leading to exhaustion or suffocation; from inflammation of the pulmonary structure or pleuræ; from irritative fever accompanying profuse discharges; or from fluid accumulations in one or both of the pleural sacs.

In gunshot injuries a penetrating wound of the chest is in most instances readily obvious to the sense of sight or touch; but it will be found by no means easy always to decide whether a lung has been penetrated or otherwise. The train of symptoms usually described as characterizing wounds of the lung must not be expected to be all constantly present; they are each liable to be modified by a great variety of circumstances, and may each severally exist in penetrating wounds of the chest where the lung has escaped perforation. Nor is it always easy to determine whether the ball has lodged or not; or, the ball having passed through, whether fragments of bone, or other substances, have remained behind.

When the chest has been opened by a projectile, the following signs may be expected in addition to the external physical evidences of the injury: a certain amount of constitutional shock; collapse from loss of blood; and, if the lung be wounded, effusion into the pleural cavity, hemoptysis, dyspnœa, and an exsanguine appearance. These will generally, but not invariably, be followed, after twenty-four hours or later, by the usual signs of inflammation in some of the structures injured.

The shock of penetrating wounds of the chest, apart from the collapse consequent on hemorrhage, is not generally so great as happens in extensive injuries to the extremities or in penetrating wounds of the abdomen. There is often much more "shock" when a ball has not penetrated; but, having met with something to oppose its course, has nevertheless inflicted a violent percussion of the whole chest and its contents.

When loss of blood occurs without the lung being wounded, the hemorrhage is probably proceeding from a wound of one of the intercostal arteries, which has been torn by the sharp ends of fractured bone. Serious hemorrhage, however, is exceedingly rare from vessels external to the cavity of the chest.

When blood is effused in any large quantity into the pleural sac—as indicated by the exsanguine appearance of the patient, increasing dyspnœa, occasional hemoptysis, and the stethoscopic signs on auscultation,—the inference is, that the lung has been opened, and that it is from its structure the blood is flowing. The amount of hemorrhage in wounds of the lungs will greatly vary according to the direction of the track of the ball; for the large vessels cannot here glide away from the action of the projectile, as they may in the neck or extremities of the body. Wounds, therefore, near the root of each lung, where the pulmonary arteries and veins are largest, are attended with the greatest amount of

hemorrhage; and as coagula can hardly form sufficiently
to suppress the flow of blood, are generally fatal.

Hemoptysis indicates injury to the lung, but does not give
assurance that this organ has been penetrated. It generally
accompanies gunshot wounds of the lung in a greater or less
degree, no doubt always when a bronchial tube of large size
is penetrated; but, as may be ascertained by careful perusal
of recorded cases, is sometimes wholly absent, even though
the patient may be troubled by cough. Dr. Fraser, in a
recent monograph on Wounds of the Chest, states that out
of nine fatal cases observed by him in the Crimea in which
the lungs were wounded, only one had hemoptysis; and out
of seven in which the lungs were found not to be wounded,
two had hemoptysis. This, however, from the writer's observation, would appear to be an unusual proportion of
cases in which hemoptysis was not present after wounds of
the lungs.

Dyspnœa is a frequent accompaniment of wounds penetrating the lung, but not a constant symptom before inflammatory action has set in. When dyspnœa is great in the early
period, it will often be found to depend upon the injuries
to the parietes, and to the pain caused on taking a full inspiration; as a sign of subsequent mischief in the progress
of the case, it is, of course, very constantly present. It is
now known that the opening of the pleura does not necessarily induce collapse of the lung, even though unfettered by
adhesions, during life. It was formerly supposed that the
escape of air by the wound was a sufficient proof that the
lung had been opened by the projectile; but it is evident
that it is not so, as the air may enter by the wound and be
forced out again by the expansion of the lung in inspiration,
or by the action of the chest on expiration. If air and frothy
mucus with blood, as noticed in one of the cases recorded in
the Crimean campaign, escape by the wound, there can be
no doubt of the nature of the injury. Emphysema is not
common in penetrating gunshot wounds, but occasionally

happens. The free opening generally made by the projectile sufficiently explains this fact.

It is not necessary to refer at any length in this place to the inflammations which may supervene. Diffused inflammation of the lung after wounds is not so common as might perhaps be expected. In unfavorable cases, the pleural cavity is generally found to be the seat of extensive inflammatory action or unhealthy accumulations, especially where irritation has been kept up by the presence of foreign bodies or the patient's constitution has become from any cause debilitated.

Treatment.—The object of the surgeon's care must be in the first place to arrest hemorrhage ; afterward, to remove pieces or jagged projections of bone, or any other sources of local irritation ; and to adopt means to prevent interference with the natural process of cure, which takes place by adhesion of the opposite pleural surfaces near the wound in the first instance, and subsequently by cicatrization of the wound itself, or, as shown in an interesting preparation in the museum of the Army Medical Department at Fort Pitt, by contraction into a narrow sinus lined with a distinct adventitious membrane into which the small bronchial tubes open. Although the shock may happen to be considerable, attempts to rally the patient, if any be made, should be conducted very cautiously ; the prolongation of the depressed condition may be valuable in enabling the injured structures to assume the necessary state for preventing hemorrhage. Hemorrhage from vessels belonging to the costal parietes should be arrested by ligature, as in other parts, if the source from which it proceeds can be ascertained, and if the flow of blood be so free as not to be controlled by the ordinary styptics. Operative interference of this kind is chiefly called for on account of secondary, not primary, hemorrhage. Hemorrhage from the lung itself must be treated on the general principles adopted in all such cases ; the application of cold to the chest, perfect quiet, the administration of opium,

and, if the patient be sufficiently strong, bleeding from a
large opening until syncope supervenes. When blood has
accumulated in any large quantity, and the patient is much
oppressed, the wound should be enlarged, if necessary, so
as, with the assistance of proper position, to facilitate its
escape. If the effused blood, from the situation of the
wound, cannot be thus evacuated, and the patient be in
danger of suffocation, then the performance of paracentesis,
as directed for the relief of empyema, must be resorted to.

The extensive bleedings formerly recommended in all penetrating gunshot wounds of the chest are now practiced with
much greater limitations—indeed, should never be employed
simply with a view to prevent mischief from arising. Venesection carried to a great extent does harm by lessening the
restorative powers of the frame. It appears to interrupt the
process of adhesion between the pleural surfaces and the
steps taken by nature to repair the existing mischief, while
it leads the injured structures into a condition favorable for
gangrene, or encourages the formation of ill-conditioned
purulent effusions. When inflammation has arisen, venesection may be joined with other means to control its excessive
action, and to give relief, which it certainly does, to the patient; and where hemorrhage is manifestly going on internally, it may be practiced with a view of draining the blood
from the system and more speedily inducing faintness, to
give an opportunity to the pulmonic vessels to become
closed; but, even when thus applied, the general state of
the patient will not be unconsidered by a judicious surgeon,
nor caution neglected, lest the venesection cause him to sink
more rapidly from the additional shock to the system and
abstraction of restorative force. Taking away blood certainly does not prevent pneumonia from supervening, but
occasionally seems to give the inflammation, when it arises,
more power over the weakened structures, or even to cause
it to be accompanied with typhoid symptoms. Many cases
will be found in the various published records derived from

the Crimean campaign, where favorable recovery has taken
place after wounds of the lung without venesection being at
all resorted to as part of the treatment.

The case of an officer of the 19th Regiment, who was
shot at the assault of the Great Redan, and under the care
of the writer, will serve to illustrate some of the points
before named. In this instance, a rifle-ball passed through
the upper part of the left scapula near its superior posterior
angle, comminuting the bone and entering the chest. The
ball, together with a piece of cloth, was excised in front,
two inches above and internal to the fold of the axilla. The
mouth was filled with blood immediately after the injury;
bloody expectoration continued for three days; there was
hacking cough on increased inspiration; the respiratory
murmur was accompanied with slight crepitating *râles* in
the upper part of the lung; there was weakness, but not
much shock. The small degree of the latter symptom, and
the absence of evidence of effusion of blood into the pleural
cavity, led at the time to a suspicion that the ball had
glanced round the costal pleura and had only contused the
lung; but the fact of the absence of vessels of large size at
this part of the lung, especially if there were pleural adhesions, may have been the cause of these results. This officer
had been much weakened in frame by scorbutic diarrhœa in
the winter of 1854–55, and though the cure was protracted
by occasional attacks of diarrhœa subsequently to the injury,
by profuse discharge from the wounds, and separation from
time to time of spiculæ of bone, he left for England two
months afterward with his recovery nearly completed, and
no inconvenience has been experienced in the discharge of
his duties since. No venesection was practiced in this
case; but tonics, nourishing diet, and port wine were given
as soon as suppurative action had been established.

But in discountenancing great bleeding, mention should
not at the same time be omitted that, in many cases, recorded by numerous authors, and judging *post factum*, the

successful issues appear to have been owing to copious
venesection. A remarkable case occurred in a young soldier
of the 33d Regiment, private Thomas Monaghan, under
the care of Deputy Inspector-General Dr. Muir, then surgeon of the regiment. This man was wounded in August,
1855, through the left shoulder-joint and chest, the glenoid
cavity and head of the humerus being injured and the lung
implicated. In this instance complete recovery as to the
chest, and recovery with partial anchylosis of the shoulder,
without operative interference, followed, and appeared attributable chiefly to inflammatory action being subdued by
repeated depletion, the use of antimonial medicines, and
enforced abstinence. In two other cases, hitherto unrecorded, which occurred during the same month in the same
regiment, successful terminations appeared to be attributable
to similar means. In one of these the ball entered the front
of the chest, between the third and fourth ribs, and passed
out between the seventh and eighth ribs below; in the other,
after passing through the right arm, it entered the chest at
the posterior border of the axilla, and emerged near the
apex of the scapula.

   To remove splinters of bone, and readjust indented portions of the ribs, the finger should be introduced into the
wound, and care taken that in doing so no pieces of cloth
or fragments be separated and projected into the pleural
sac. Notice must at the same time be taken of any bleeding
vessel requiring to be secured. A pledget of lint should be
laid over the wound, and a broad bandage placed round the
chest, just tight enough to support the ribs and in some
degree to restrain their movements, but with an opening
over each wound large enough to permit the ready access of
the surgeon to it if necessary. If the patient's comfort admits of it, he should be laid with the wound downward, with
a view to prevent accumulation of fluid in the pleura; and
if there be two openings, as will be most frequently the case
in rifle-ball wounds, one wound should be thus placed, and

the upper one kept covered. In gunshot wounds, closure of the parietes by adhesion is of course not to be looked for. The diet, beverages, and medicines must constantly have reference to the avoidance of inflammatory action; and should this occur it must be combated on general principles. It is by such means we shall best assist the natural efforts toward recovery.

If the presence of a ball within the cavity be ascertained, efforts should be made for its removal. But any attempt to determine where the ball has lodged should be made very cautiously, as more harm may result from the interference than from the lodgment of the foreign body. The existence of old adhesions will modify the effects of a penetrating wound, by excluding the track of the ball from the general pleural cavity, and may influence the result of the injury, especially if there be hemorrhage, or lodgment of foreign bodies, which may thus be brought within the sphere of removal more readily.

Wounds of the heart seldom come to the military surgeon's notice, as they ordinarily prove fatal on the battlefield. Still it is right to mention, that examples occur in which musket-balls are lodged in the heart without immediately fatal results; and one case is recorded, where a ball was found imbedded in its substance six years after the injury was received, and death then ensued from causes unconnected with the wound.* Cicatrices have also been discovered, showing that a portion of this organ had been wounded with recovery. A private of the 2d Foot, wounded in the chest, came to England in a transport, and died sixteen days afterward in the military hospital at Plymouth. On removing the heart, a ball was found in the pericardium. There was a transverse opening in the right ventricle, near the origin of the pulmonary artery, and the appearances led to the supposition that the ball had, previous to death, been

* Dict. des Sciences Méd., Paris, 1813, p. 217.

lying in the right auricle. There was general inflammation
of the heart and left side of the chest, but no signs of inflammation on the right side. A preparation of this heart is
preserved.* These are only referred to as indications of
what cases may occur among chest injuries; such accidents
are so rare as to lead to little practical result.

GUNSHOT WOUNDS OF THE NECK.

Gunshot wounds of this region do not appear to be so
fatal as might be anticipated from the large vessels and important canals leading to the thorax and abdomen, which at
first sight appear to be so exposed and unprotected. In no
region are so many examples offered of large vessels meeting but escaping from balls in their passage as in this;
because the cause which operates elsewhere—ready mobility
among long and yielding structures—exists in a greater
degree in the neck than in any other part. Where the
large vessels happen to be divided, death must follow almost
immediately.

Superficial wounds of the neck offer no peculiarities.
The larynx and trachea being the organs most prominent,
and most frequently injured, are those which chiefly attract
the surgeon's notice in warfare; but a consideration of the
anatomical structure will at once show what numerous other
complications, whether from direct injury or consequent
inflammation, projectiles are likely to cause when driven
deeply into or perforating this region.

A brief abstract of some wounds of the neck, which
occurred during the Crimean campaign, will serve to exhibit the leading symptoms connected with them when the
larynx, or larynx and œsophagus, are involved. Four cases
may be found in the *Lancet* of January 19th, 1856, to which

---

* See Edin. Med. and Surg. Journal, vol. xiv.—Case of gunshot
wound of the heart, by J. Fuge, Esq.

journal they were communicated by the late Mr. Guthrie, as "very interesting." In the Surgical History of the War it is stated that only three wounds of the neck, other than simple flesh wounds, occurred among the officers, from the commencement to the end of the war; of which two proved fatal, and one led to invaliding. The case of an officer of the 19th Regiment, however, fell under the care of the writer, which is not included in that number; and in this instance the neck was completely traversed, the œsophagus perforated from side to side, and the larynx injured. It is detailed among the cases by Mr. Guthrie. After the shock had subsided, the leading symptoms were aphonia, dysphagia, numbness of one arm, edema and stiffness of the neck, distressing accumulation of mucus about the fauces, and slight pyrexia. Recovery progressed favorably, and on the twenty-second day after the injury both external wounds in the neck were healed, and the two in the œsophagus appeared to be closed also. The patient referred to still suffers from a certain amount of aphonia, but not enough to prevent him from performing his duties as a captain, though want of sufficient power of voice would probably disable him for a more extensive command. Another of these cases, in which emphysema of the neck, edema of the glottis, great dyspnœa, and threatened suffocation gradually supervened in a superficial gunshot wound of the neck, with fracture of the thyroid cartilage, is related by Assistant-Surgeon Cowan, 55th Regiment, who performed tracheotomy, and thereby saved the patient's life. In another, the ball passed through the thyro-hyoid membrane, fractured the thyroid cartilage, and tore the lining membrane of the glottis. Tracheotomy was performed on the day after the injury, without benefit. Liquids could not be prevented from passing into the trachea through the wound made by the projectile. The fourth case above referred to was in a private of the 97th Regiment. The ball entered at the pomum Adami, and passed out by the anterior edge of the right sterno-mastoid muscle. Loss of voice, frequent cough, bloody sputa, slight emphysema at the wound of entrance, and nausea, were the leading symptoms. When the man attempted to drink, some of the fluid escaped by the wound of exit. After five days this occurrence ceased; and after the twelfth day, air no longer passed out of the wound of entrance. Both wounds gradually healed; but aphonia—the voice being reduced to a whisper —existed when the man left the regimental hospital. A soldier of the Rifle Brigade, under the care of Deputy Inspector-General Fraser, C.B., then surgeon of the battalion, was shot through the trachea, and respiration was for some time carried on by the wound; it, however, gradually and completely healed, and a favorable recovery ensued. Another interesting case, hitherto unrecorded, occurred in a soldier of the same battalion, at the last assault of the Redan. A rifle-ball entered this man's neck at the lower part of the left sterno-mastoid muscle, passed across under the skin, wounding the anterior surface of the trachea, severed some fibers of the right sterno-mastoid, and effected its exit. The man was wounded at the same time by two other rifle-balls, both flesh wounds, one through the left forearm, the other through the upper part of the right thigh; while a shell exploding near him, caused his left eye to be penetrated with particles of stone and earth. Vision was lost; but in other respects, excepting a little lameness from the wound in the thigh, he was discharged cured, after fifty-six days' hospital treatment.

Seven cases of gunshot wounds of the neck returned to England from the late mutiny in India. They were all simple flesh wounds. In one the musket-ball had not been discovered, and its position remained unknown. The man was wounded at Lucknow, and the ball entered the left side of the neck, close to the thyroid cartilage. Baron Percy reports a similar wound and case of lodgment in his *Army Surgeon's Manual;* in this instance, the ball was known to pass away by the bowels, a fortnight after the injury was received.

The liability to concussion of the cervical portion of the vertebral column, and to injury of the deep cervical and other nerves, must not be overlooked. Wounds of the neck are often accompanied by more or less loss of power in one of the upper extremities; and more extensive paralysis occasionally succeeds, although there was no primary evidence of the spine being implicated in the injury.

GUNSHOT WOUNDS OF THE ABDOMEN.

Gunshot wounds of the abdomen, like those of the chest, are, for the sake of convenience, divided into *non-penetrating* and *penetrating*. The NON-PENETRATING may be either simple flesh wounds, or may be accompanied with fracture of some of the pelvic bones, or with injury to some of the contained viscera. In PENETRATING wounds, the peritoneum only, or, together with it, one or more of the abdominal viscera, may be wounded; or, in comparatively rare cases, a viscus may be penetrated without the peritoneum being involved. It is in the regional cavity of the abdomen that the proportion of penetrating wounds is the greatest. The cranium, from its form, structure, and coverings, serves as a strong defense even against gunshot; the osseous yet elastic and movable ribs, the sternum, and muscular parietes greatly protect the contents of the cavity which they inclose; but the extensively exposed surface of the abdomen, anteriorly and laterally, has no power of resistance to offer against a projectile directly impinging it; and when this important cavity is once penetrated by these means, death is the almost inevitable result. Even the chances of a favorable termination which may exist in wounds from other causes are generally wanting; and much of their treatment, such as the use of sutures, and other means to insure the apposition of cut edges, is inapplicable, from the parts to a certain distance being almost necessarily deprived of their vitality, to injuries from gunshot.

Non-penetrating wounds require but few remarks in this place. The fatal injuries which occasionally occur from masses of shell or round shot, in which the liver, spleen, or other viscera are ruptured without penetration of parietes, and where death ensues from shock, hemorrhage, or peritonitis, have already been alluded to. If, although the viscera have been contused, the injury does not amount to being mortal, the patient should be subjected to perfect quiet, extreme abstinence, and, only when inflammation arises, to the necessary treatment for its control. If the parietes have been much contused, abscess or sloughing may be expected; and a tendency to visceral protrusion must be afterward guarded against.

When portions of the pelvic parietes are fractured by heavy projectiles, very protracted abscesses generally arise, connected with necrosed bone; and the vital powers of the patient are greatly tried by the necessary restraint and long confinement. The great force by which these wounds must be produced, and the general contusion of the surrounding structures, cause a large proportion sooner or later to prove fatal, notwithstanding the peritoneal cavity may have escaped. Of twenty-nine such cases which came under treatment in the Crimea, sixteen died. Even apparently slight cases, as where a portion of the crest of the ilium is carried away by shell, or ball lodged in one of the pelvic bones, often prove very tedious, from the long-continued exfoliations and abscesses which result.

Penetrating wounds.—A penetrating wound of the abdomen, whether viscera be wounded or not, is usually attended with a great amount of "shock." The prognosis will be extremely unfavorable, if there is reason to fear the projectile has lodged in the cavity of the peritoneum; and in all cases the danger will be very great from inflammation of this serous investment. The liability to accumulation of blood in the cavity, from some vessel of the abdominal wall being involved in the wound, must not be forgotten.

When, in addition to the cavity being opened, viscera are penetrated, and death does not directly ensue from rupture of some of the larger arteries, the shock is not only very severe, but the collapse attending it is seldom recovered from up to the time of the fatal termination of the case. This is sometimes the only symptom which will enable the surgeon to diagnose that viscera are perforated. The mind remains clear; but the prostration, oppressive anxiety, and restlessness are intense; and, as peritonitis supervenes, pain, dyspnœa, diffused tenderness, irritability of the stomach, distention, and the other signs of this inflammation are superadded. In ordinary wounds from musket-shot, scarcely any matter will escape from the opening of the parietes, the margin of which becomes quickly tumefied; but if any escape, it will probably indicate what viscus has been wounded. If the stomach has been penetrated, there will probably be vomiting of blood from the first. If the spleen or liver be wounded, death from hemorrhage is likely to follow quickly. In some instances patients, however, recover after gunshot wounds involving these viscera, and examples in illustration may be found in various works on military surgery. Two particularly manifest instances, where officers were shot through the liver by musket-balls, occurred lately in India; one at Lucknow, the other at the siege of Delhi: both recovered. The cases are described in the *Indian Annals of Medical Science* for January, 1859. If the small intestines have been perforated, and death follows soon after from peritonitis, the bowels usually remain unmoved, so that no evidence is offered of the nature of the wound from evacuations; but in any case of penetrating wound of the abdomen, when the opportunity is offered, steps should be taken—a matter not unlikely to be omitted under the circumstances of camp hospitals full of patients—to isolate and examine all evacuations which may follow. By attending to this direction, the writer had the satisfaction of ascertaining the passage of a ball and piece of cloth, after a wound in the loin, in a case already alluded to. If the kidneys or bladder are penetrated, the escape of urine into the abdomen is almost a certain cause of fatal result. The latter viscus may, however, be penetrated without the peritoneal cavity being opened; and, as experience proves, the wound is then by no means of a fatal character. Musket-balls sometimes lodge in the bladder. This was ascertained to have happened in a soldier of the 20th Regiment, in the Crimea; but the patient died from other injuries, so that the information could not be turned to account. Mr. Guthrie performed the usual operation of lithotomy, with success, to remove a musket-ball which had struck a soldier just above the pubes, at Waterloo, and lodged. He also records a similarly successful case in a man wounded at the battle of Chillianwallah: this ball formed the nucleus of a calculus. Baron Percy removed a ball and a portion of shirt from the bladder. In all such cases, it is probable that the bladder has been penetrated at some part uncovered by peritoneum, so that the cavity of the abdomen has not been opened; or, if otherwise, the foreign body has found its way in by ulceration, after adhesions had been established, and thus circumscribed the openings of communication. Small foreign bodies may also pass into the bladder by the ureter. A case in which the kidney was wounded came under the care of the writer, after the 8th of September, 1855. The patient survived twelve days, and then died from pyemia. He had been taken prisoner, but was found in Sebastopol, and brought to his regimental hospital on the second day after the assault. There was only one wound in the right loin, and the ball had lodged. Extensive abscesses formed among the gluteal muscles on the left side, and down the left thigh; and though free incisions were made, great constitutional irritation supervened, and he sank. The substance of the right kidney had been perforated, but the ureter had escaped. The ball had passed across the abdomen, and lodged in the left buttock. Mr. Guthrie mentions some wounds of the kidney where recovery took place; in one, seven months after the wound, after an attack of retention of urine, a piece of cloth was forced out by the urethra, which must have come down from the pelvis of the kidney. When the abdominal parietes have been opened by shell or passage of large shot, protrusion of omentum and intestines will probably be one of the results. This does not always happen. In Dr. Macleod's Notes, p. 237, is detailed a remarkable case of recovery, which was witnessed by the writer, after the wall of the abdomen, including the peritoneum, had been destroyed to the extent of five inches long by three broad; and a coil of intestine laid bare without protrusion, in the right iliac region. This patient had also a fracture of the ileum, another of the great trochanter on the same side, and his right forearm smashed. This case was treated in the general hospital before Sebastopol, by Mr. Rooke. Sometimes a wound caused by a large projectile, which was at first not penetrating, will indirectly become so, from the severe contusion and consequent sloughing to such an extent as to denude the viscera; and if, as is not unlikely, adhesion has taken place in the mean time between a portion of the viscera and peritoneal lining of the abdominal paries, the sloughing action may extend more deeply and the bowel itself become opened.

Curious instances are recorded in which balls have passed directly through the abdomen without perforating any important viscus, as proved by examination after death. As an example, on the other hand, of the number of wounds which may thus be inflicted, a soldier of the 19th Regiment, on duty in the trenches before Sebastopol, who was shot through the abdomen in the act of defecation, was found by the writer, on post-mortem examination, to have had as many as sixteen openings made in the small intestine. He survived the wound nineteen hours.

Gunshot wounds of the colon, especially of the sigmoid flexure, appear to be less fatal, probably from structural causes as well as circumstances of position, than wounds of the small intestine. In the Museum of Fort Pitt, however, is a preparation of jejunum exhibiting three constrictions, and supposed to have been perforated in three places, from a private of the 80th Regiment, who was shot through the abdomen at Ferozeshah, in 1845, and who died from cholera in 1851. Inspector-General Taylor, C.B., then surgeon of the regiment, who made the examination post mortem, thus described the injured part of the intestine : " The intestines neither there nor elsewhere were morbidly adherent; but the fold of intestines immediately opposed to the cicatrix presented a line of contraction as if a ligature had been tied round the gut. The same appearance existed in two other places." It seems more likely that the gut was contused than perforated, and that contraction gradually supervened, especially as no adhesions were found ; and, when wounded, the symptoms were so slight as to have led to the supposition that the ball had gone round the abdominal wall.

A gunshot wound of the intestine, more especially the colon, may lead to fecal fistula, and life be thus saved for a time. One such case only occurred in the Crimea, in the 19th Regiment, of which the writer was then the surgeon ; this case, which has been before casually mentioned, subsequently passed under the care of his friend Mr. Birkett, of Guy's Hospital, in which institution the patient died, from the effects of albuminuria, four years after the receipt of the wound referred to. The surgical history of this case has been already published at some length in the *Lancet;** the medical history, together with the results of the post-mortem inspection, have been detailed by Dr. Habershon, in vol. v., Ser. III., of the *Guy's Hospital Reports.* The fistula became closed at intervals, and occasionally, before other dis-

---

* For 1855, vol. i. p. 606, and vol. ii. p. 437.

ease supervened, hopes were entertained that recovery might result. The direction and depth of the wound precluded any of the usual operations for attempting to effect a radical cure. Two cases of abnormal anus by gunshot perforation are recorded by Dr. Williamson among the wounded who have recently returned from India; in both instances the descending colon was the part of the bowel implicated. A similar result is recorded in a private of the 13th Regiment wounded at Cabul in 1840.

Wounds of the diaphragm.—Musket-balls occasionally pass through the diaphragm; and Mr. Guthrie has remarked that these wounds, in instances where the patients survive, only become closed under rare and particular circumstances. Hence the danger of portions of some of the viscera of the abdomen, as the stomach or colon, passing into the chest, and thus forming diaphragmatic herniæ, and of these, eventually, from some cause becoming strangulated. Two very interesting preparations of these accidents from gunshot exist in the museum at Fort Pitt. In both instances, the stomach, colon, and omentum form the hernial protrusions. In one, death occurred, a year after the wound, from strangulation induced suddenly after a full meal; in the other, the soldier continued at duty twenty-two years after, and died from other causes. All the cases which occurred in the Crimea in which openings had thus been established between the cavities of the chest and abdomen proved fatal. A case is detailed in the Surgical History of the War where the patient survived a double perforation of the diaphragm, together with a wound of the liver, six days; in another instance, where the lung, diaphragm, liver, and spleen were wounded, the soldier lived sixteen hours. The direction of the ball, hiccough, dyspnœa accompanied with spasmodic inspiration, and inflammatory signs more particularly connected with the chest will be the usual indications of such a wound; and in case of recovery, the risk of hernial protrusion and strangulation must be explained to the patient.

Should strangulation occur, it can hardly be expected that division of the stricture could be performed without the operation itself leading to equally certain fatal results.

Treatment. — In the general treatment of penetrating wounds of the abdomen by gunshot, the surgeon can do little more than to soothe and relieve the patient by the administration of opiates, and to treat symptoms of inflammation when they arise on the same principles as in all other cases. The usual directions to attempt agglutination of the opposite portions of peritoneum by favorable posture cannot generally be carried out, the attempts being defeated by the restlessness of the patient. The collapse which attends such injuries may be useful in checking hemorrhage; and the exhibition of stimulants is further contra-indicated by the risk of exciting too much reaction, should the wound not prove directly fatal. If the wound be caused by grape-shot or a piece of shell, and intestine protrudes, it must be returned; if the intestine be wounded, sutures are inapplicable, as in an incised wound, without previously removing the contused edges. When the bladder is penetrated, care must be taken to provide for the removal of the urine, either by an elastic cathether, or, if this cannot be retained, by perineal incision. A freely communicating external wound prevents the employment of the catheter from being essential. A soldier of the 57th Regiment was wounded, on the 15th June, 1855, by a musket-ball, which entered the left buttock, fractured the pelvis, and came out about three inches above the os pubis and one inch to the right of the median line. The bladder was perforated; urine escaped by both openings, chiefly by the one in front. Here the catheter caused so much irritation that it was withdrawn; but the posterior wound soon ceased to discharge urine, and in eighteen days the anterior wound was free from discharge also. Seven weeks after the date of injury symptoms resembling those of stone in the bladder came on; these were relieved on three spiculæ of bone making their escape by the urethra. About the same time the anterior wound became again open, and some pieces of bone were discharged. After ninety-seven days' treatment in the Crimea, the man was sent home—the anterior wound being still so far open that distention of the bladder, as from accumulation at night-time, led to a little oozing from it. This subsequently healed; and he was sent to duty on the 22d of November, nearly six months after the date of injury.

GUNSHOT WOUNDS OF THE PERINEUM AND GENITO-URINARY ORGANS.

From the position of these parts of the body, uncomplicated gunshot wounds of them are comparatively rare. Throughout the whole of the Crimean war, the number of cases treated amounted, among the men, to 70; among the officers, only to 4. The number of deaths which resulted were 21 among the men, chiefly cases of extensive laceration involving the urinary apparatus; among the officers, none. Three men only, out of 603 who returned from the late mutiny in India to Chatham, are recorded under this class. In one, the injury was from a spent shot, which caused a bruise without laceration over the symphysis pubis, and produced persistent incontinence of urine; in each of the other two, a musket-ball wounded the left testicle, injured the urethra, and led to urinary fistula, which was, however, afterward healed. In one, the testicle was so much injured that it was removed on the day the wound was received; in the other, it sloughed away shortly after. A corporal of the 19th Regiment, wounded in this region on the 8th September, 1855, was under the care of the writer. A portion of the ascending ramus of the ischium on the right side was driven into the perineum, the soft parts were much injured, and the right testicle was destroyed. The viscera of the pelvis escaped. He was doing well until nearly a fortnight after the injury, when nervous irritation and trismus set in, and he sank.

Perineal wounds are not unfrequently caused by shells bursting and projecting fragments upward; but they are generally mixed with lesions of viscera of the pelvis, or fracture of its structure, or injuries about the upper parts of the thighs or buttocks. In one such case, a portion of the scrotum, the whole of one testicle, and the greater part of the other were carried away. This wound healed without fungous growth from the remaining portion of the testis. Separate wounds of the external organs of generation are usually caused by bullets. In two cases in the Crimea, a bullet entered between the glans penis and prepuce, and traversed upward without penetrating the erectile tissue. M. Appia records a case where the ball entered the summit of the glans, traversed the whole length of the corpus cavernosum, passed under the pubic arch, and went out by the right buttock. The urethra was not opened. Double orchitis and scrotal abscesses followed; but favorable cure took place. In another case, a ball carried away the inferior part of the glans but did not wound the urethra. A soldier of the Rifle Brigade was wounded in the Crimea by a musket-ball, which entered the right buttock and came out by the body of the penis, just below the glans, having ruptured the urethra about four inches from the meatus. The wound of the penis closed favorably. Mr. C. Hutchinson has recorded the case of a soldier of the 42d Regiment, treated at the Deal Naval Hospital, who was wounded in the upper part of the thigh by a musket-ball, which lodged. Three weeks afterward, the ball was found imbedded in the pubes, the urethra being stretched around the convex surface; and this explained the cause of a distressing distention of the penis and dribbling of urine which had existed without intermission from the time of the injury, but ceased at once on the removal of the bullet.

GUNSHOT WOUNDS OF THE EXTREMITIES.

These injuries, always very numerous in warfare, offer many subjects of consideration for the military surgeon. No class of wounds includes so many cases that fall under his prolonged care as this. A large proportion of wounds of the head and trunk are immediately fatal, or from the com-mencement contain the elements of fatal results; while wounds of the extremities, if those of the thigh be excepted, are free from this extremely serious character. The treat-ment to be pursued, including questions of conservation, re-section, amputation, and the proper time for the adoption of these latter if determined upon, often demands the closest attention of the surgeon. These subjects will be considered in their general bearing in other parts of this work, and only those points especially connected with the circumstances of warfare will be here referred to.

Gunshot wounds of the extremities divide themselves into flesh wounds and contusions, and those complicated with frac-ture of one or more bones. Flesh wounds may be simple, and these offer few peculiarities, whatever their site; or they may be accompanied with lesion to nerves, or blood-vessels, or both, and these usually increase in gravity in proportion as they approach the trunk.

When complicated with fracture, the lesion is usually ren-dered compound by the direct contact of the projectile with the bone injured; but the fracture is sometimes simple, when caused by indirect projectiles, such as stones or splin-ters, or by spent balls. These injuries are liable to become further aggravated by the fracture extending into or being complicated with an opening of one of the joints. Joints may be contused or opened by projectiles, without apparent lesion of any portion of the bones entering into their com-position; but these are exceptions to the usual order of such cases from gunshot.

Simple flesh wounds have already been referred to both
in respect to their nature and treatment in the commencement of this essay. It is in connection with fractures of
bones and their proper treatment that the interest of surgeons is chiefly attracted in gunshot wounds of the extremities. From the nature of the injuries, already described, to
which bones are subjected by the modern weapons of war,
together with the irreparable nature of the wound in the
softer structures, except after a long process of suppuration
and granulation, as well as from the usual circumstances of
military life, it might be anticipated that difficulty would
often arise in determining which of the double set of risks
and evils—those attending amputation, and those connected
with attempts to preserve the limb with a profitable result—
would be least likely to prove disadvantageous to the patient.
Experience in such injuries has established certain rules
which are now generally acted upon; some still remain *sub
judice*.

Although the subject of pyemia is considered in its general
bearings elsewhere, it is right to mention here that this
serious complication, as met with in gunshot wounds, appears to be especially induced by injuries of bones, particularly those of long bones in which the medullary canal has
been laid open and extensively splintered. Several circumstances probably conduce to this result: the prolonged suppurative action during the removal of sequestra, the irritation caused by sharp points and edges, sometimes increased
by transport from primary to secondary hospitals, the patu-
lous condition of veins in bones leading to thrombosis, being
its chief local sources; while depressed vital power from any
cause, and continued exposure to an impure atmosphere
from the congregation of numerous patients with suppurating wounds, are the principal agents in producing the state
of constitution favorable to its development and progress.
Unless the hospital miasmata engendered in this way are
constantly removed as they arise, or very greatly diluted by

proper ventilation, it is almost impossible that patients
laboring under severe wounds of the extremities with comminuted bony fractures can be long saved from septicemia
and pyemia; and these, when they supervene, rarely lead to
any but a fatal termination. The different conditions of
hospital air, which in one set of cases lead to the appearance
of hospital gangrene, in another set of pyemia, are not
properly understood; but from the frequency with which
the latter complication follows wounds of bones, it would
seem that an especial influence is exerted by the local peculiarities of these injuries already mentioned. However, observation would also lead to the belief that certain individuals
are much more predisposed to pyemic action than others
placed under similar circumstances. Occasionally, in gunshot injuries of bones, where no splintering has occurred,
but only a small portion of the periosteum has been torn off
and the shaft contused by the stroke of a bullet, severe inflammation will follow, the medullary canal become filled
with pus, and death ensue from pyemia. The attention of
surgeons has been particularly called to the various circumstances producing inflammation and suppuration of the
medullary tissues—osteo-myelitis—in long bones after gunshot injuries by M. Jules Roux of Toulon.*

Upper Extremity.—Fractures of the bones of the arm
are well known to be very much less dangerous than like
injuries in the corresponding bones of the lower extremity.
Unless extremely injured by a massive projectile, or longitudinal comminution exist to a great extent, especially if
also involving a joint, or the state of the patient's health be
very unfavorable, attempts should always be made to preserve the upper extremity after a gunshot wound. In the
Director-General's History of the Crimean Campaign, the

---

* Bulletin de l'Académie Impériale de Médecine, 24th April, 1860.
See also Des Amputations consécutives à l'Ostéomyélite dans les Fractures des Membres par armes à feu, par M. H. Baron Larrey, Paris,
1860.

10

recoveries without amputation are shown to be, in the humerus, 26·6; radius and ulna, 35·0; radius only, 70·0; ulna only, 70·0 per cent. of cases treated. The proportion of deaths in these cases was only 2·3 per cent. Although not the result of gunshot, a remarkable case, published by Staff-Surgeon Dr. Williamson, by whom the operation was performed, serves to illustrate how extensively bone may be removed from the upper arm, and a useful member be still retained. The details will be found in his Notes on the Wounded from the Mutiny in India. The whole of the ulna, (not merely sequestra, but also the new bone which had formed around them, the object of which proceeding is not stated,) two inches of the humerus, and the head and neck of the radius were removed; and, four months after the operation, the man could " bend his forearm, raise his hand behind his head, lift a 28-lb. weight from the ground, pronate and supinate the hand, and use his fingers well." Of 194 wounds and injuries of the upper extremity among men returned from the late mutiny in India, 100 are recorded by Dr. Williamson to have been sent to duty regular or modified, 67 invalided from the service, 1 died, and 26 were still under treatment.

In the latter part of the Crimean campaign, when the health of the troops and means of treatment were favorable, it was often remarkable what extensive injuries of the upper extremity, even where the joints were involved, were repaired without amputation. The following cases are examples: Sergeant Bacon, 7th Fusileers, aged thirty-six, at the attack on the Redan on the 8th of September, 1855, was wounded by a rifle-ball, which entered the head of the left humerus, shattered the bone very much, and was extracted from below the left scapula. Dr. Moorhead determined to try to preserve the limb. The head of the humerus required to be removed in small, broken fragments; and the shaft, being found to be split down between three and four inches, was to that distance removed by the saw. The case progressed favorably, and in 1857 this man was in London with a most useful arm. A young soldier of the 23d Regiment was wounded, on the 15th August, 1855, by a large grape-shot, which passed through the right arm near the shoulder, comminuting the bone for three inches and extensively destroying the soft parts. Staff-Surgeon Williams, in medical charge, despairing of saving the limb, proposed to amputate, but, at the suggestion of the late Director-General Alexander, then principal medical officer of the Light Division, arranged to allow some days to elapse to watch symptoms. The case progressed so well that the idea of amputation was abandoned, and the man recovered with a very serviceable arm. In another regiment of the Light Division, the 77th, a healthy young soldier, under the care of Surgeon Franklin, was wounded at the last assault of the Redan, and sustained a comminuted fracture of the humerus, had the elbow-joint opened, both bones of the forearm broken about two inches below the joint, and the soft parts widely opened, by a piece of shell. Here no excision was practiced, but fragments removed as they became loose ; the arm, with its dressings, was supported on a zinc-wire cradle, hollowed out and bent at the elbow to the desired angle ; and nourishment, with malt liquor, were freely given from the first day. Anchylosis was established, and he left for England with a useful limb. The fractures above and below the joint prevented the application of passive motion.

In these injuries, where the bone is much splintered, the detached portions, and any fragments which are only retained by very partial periosteal connections, should be removed ; projecting spiculæ sawn or cut off ;* the wound being ex-

---

* Dupuytren made a division of the splinters of bone broken by gunshot into three classes, viz. : primary sequestra, those directly and completely separated by the force of the projectile; secondary sequestra, those retaining partial connections by periosteal, muscular, or other attachments, but afterward thrown off during the suppurative process; and tertiary sequestra, or necrosed portions, pro-

tended at the most dependent opening where two exist, or
fresh incisions being made for this purpose, if necessary;
light water-dressing applied; the limb properly supported;
and the case proceeded with as in cases of compound fracture from other causes. (See FRACTURE.) The same general
rules also apply in preserving as much of the hand as possible, in gunshot injuries.   If the shoulder or elbow joint be
much injured, but the principal vessels have escaped, the
articulating surfaces and broken portions should be excised.
Care should be taken to see that the projectile has wholly
passed out, or been removed.   In a case of comminuted
fracture of the humerus, in the 88th Regiment, no union
having taken place a month after the injury, and some
dead bone requiring removal, an incision was made for this
purpose, when half the bullet was found between the fractured ends.   Good union, with free motion of the arm, resulted, after this foreign body and the necrosed bone were
taken away.   The results of excision practiced in the shoul-

duced by the effects of the contusion and prolonged inflammatory
action in parts adjoining the seat of fracture. In accordance with
this arrangement, the removal by the surgeon of the primary and
secondary splinters has been regarded as simply anticipating nature
in her work; but Dr. Esmarch states, as one result of the experience
of the surgeons of the Sleswick-Holstein army, that, in the majority
of comminuted fractures, the removal of splinters retaining any connection with periosteum is unnecessary and often injurious, as is also
the practice of sawing off the broken ends of the bone projecting
from the comminuted part. By proper treatment and under favorable circumstances, he asserts, such splinters become impacted in
callus, and in time unite with the other fragments of the bone, and
in this manner a cure is completed without operative interference.
It is a matter, however, of frequent observation that splinters which
have thus become impacted in callus lead to mischief in various ways,
or are subsequently discharged as if they were so many foreign
bodies, while the removal of the jagged ends of the broken bone
seems to be a valuable means of preventing irritation, and thus of
favoring union between them; and English surgeons, therefore,
generally pursue the practice above recommended.

der and elbow joints, especially the former, after gunshot wounds, have been exceedingly satisfactory. Especial attention was directed to the practice of resections of joints after gunshot injuries in the Sleswick-Holstein campaigns between 1848 and 1851; and Dr. Friedrich Esmarch has published the results in a valuable essay on the subject. Of nineteen patients in whom the shoulder-joint was resected, in twelve a more or less useful arm was preserved; and seven died. Complete anchylosis did not occur in any one instance; and in several the power of motion became so great as to enable the men to perform heavy work. Of forty patients for whom resection of the elbow-joint was performed six died, thirty-two recovered with a more or less useful arm, one remained unhealed at the time Dr. Esmarch wrote, (1851,) and in one mortification ensued and amputation was performed. These operations present no peculiarities in the mode of performance or their after-treatment, as compared with similar resections in civil practice.

Lower extremity.—Gunshot wounds of the lower extremity vary much more greatly in the gravity of their results, as well as in the treatment to be adopted, according to the part of the limb injured, than happens in those of the upper extremity. As a general rule, ordinary fractures below the knee, from rifle-balls, should never cause primary amputation; while, excepting in certain special cases, in fractures above the knee, from rifle-balls, amputation is held by most military surgeons to be a necessary measure. The special cases are gunshot fractures of the upper third of the femur, especially where the hip-joint is implicated; for in these the danger attending amputation itself is so great that the question is still open, whether the safety of the patient is best consulted by excision of the injured portion of the femur, by simple removal of detached fragments and trusting to natural efforts for union, or by resorting to amputation. The decision of the surgeon must generally rest upon the extent of injury to the surrounding structures, the

10*

condition of the patient, and other circumstances of each particular case. If the femoral artery and vein have been lacerated, any attempt to preserve the limb will certainly prove fatal.

The femur — the earliest formed, the longest, most powerful, and most compact in structure of all the long bones of the body—can only be shattered by a ball striking it with immense force. Attention was specially directed in the late Crimean campaign to the question of the proper treatment of these injuries, and expectations were generally held that the advanced experience in conservative surgery would lead to many such cases terminating favorably with preservation of the limb, which previously would have been subjected to amputation. Toward the latter part of the war, all the circumstances of the patients were as favorable for testing this practice as they have been in the various *émeutes* in Paris, with the advantages of immediate attention and all the appliances of the best hospitals close at hand. Yet, in the Surgical History of the Campaign, it is stated that only fourteen out of 174 cases of compound fracture of the femur among the men, and five out of twenty among the officers recovered without amputation being performed; that those selected for the experiment of preserving the limb were patients where the amount of injury done to the bone and soft parts was comparatively small; that where recovery ensued, it always proved tedious, and the risks during a long course of treatment numerous and grave; and that the proportion of recoveries would not appear even so large as the above, if the deaths of those who after long treatment were subjected to amputation as a last resource were included. Amputations of the thigh, however, were very fatal in their results also, the recoveries being stated to be, among the men, in the upper third 12_1^9, in the middle third 40, in the lower third $43\frac{3}{10}$, per cent. of cases treated. Among the officers the proportion was rather more favorable. But this percentage includes those cases in which attempts had been made to preserve the limb, and failure resulting, amputation was resorted to as a last chance of saving the patient, so that they ought to have been excluded from the lists of amputations, both primary and secondary, as commonly interpreted. On account of this comparatively indifferent success of amputation, resection of portions of the shaft of the femur was sometimes practiced; but the records state that no success attended the experiment, every case, without exception, having proved fatal.

In considering the results of gunshot fractures of the femur, the situation of the injury is a matter of great importance, whether as regards chances of recovery without or with amputation. In the Surgical History of the Crimean Campaign this fact is shown in the results of amputation; but the distinction is not made in regard to the recoveries without amputation. Dr. Macleod, in his Notes, remarks that he has only been able to discover three cases in which recovery followed a compound fracture in the upper third of the femur without amputation: one, that of an officer of the 17th Regiment; the second, of a soldier of the 62d; and a third, whose regiment is not named. A case, however, was under the care of the writer, not included in the above, nor appearing in the official history of the war; and one, judging from the results described in Dr. Macleod's Notes, more fortunate in its issue than at least two of the number he mentions. With regard to the first patient, Dr. Macleod states he has been informed "that although his limb was in a very good condition when he left for England, the trouble it has since given him, and the deformed condition in which it remains, makes it by no means an agreeable appendage;"* in the second, the fracture was in the lower part of the

* The officer referred to must have greatly improved in condition since Dr. Macleod wrote, as he has been of late on active service in India.

upper third, and the injury was comparatively slight; in the third, a mass of callus was thrown out which connected the bone, but he died of purulent poisoning, and never left the Crimea. In the case which was under the writer, the fracture was within the upper third; there is no distortion, and shortening only of $1\frac{1}{2}$ inches; the officer is able to walk or ride without any inconvenience, and competent for all duty. All the circumstances were most favorable for recovery in this instance; and a consideration of these on the one hand, and the experience of the unfavorable results of amputation in this region on the other, led to the effort to save the limb. A short history of this case will be useful. Lieutenant D. M., 19th Regiment, aged seventeen, of sanguine temperament, healthy frame, was brought up to camp about 4 A.M. Sept. 9th, 1855. He had been wounded in the assault upon the Redan in the upper part of the left thigh, and had been lying by the side of the ditch where he fell thirteen hours. When discovered, he was carried carefully in a soldier's greatcoat as far as the opening of the trenches, and thence on a stretcher to camp. He was very cold and prostrate on his arrival. The wound in his left thigh had been caused by a ball, which had passed out. It entered posteriorly at the fold between the left nates and thigh, three inches from the tuberosity of the ischium; passed forward, downward, and outward, and made its exit seven inches below the trochanter major. The femur was broken in the line of passage of the ball, which, from entrance to exit, appeared to be about six inches. From the trochanter major to the seat of fracture was four inches; to the external condyle on the same side was $15\frac{1}{2}$ inches. The amount of comminution appeared slight, but, from its vicinity to the joint, the great swelling about the limb, and desire to avoid aggravating pain, the precise condition of fracture was not further ascertained. The upper fragment projected forward, but any attempts at reduction caused great suffering; and some restoratives being given, wet compresses applied to the thigh, and the limb secured against additional movement, the patient was left to rest. At a consultation the following morning, from the patient's age, so favorable for reparative action, very healthy constitution, and the fact that, the siege being over, full attention could be paid to the case, conservation of the limb was settled to be attempted, and the patient was therefore treated with this view. In addition to the wound just named, he had received an extensive contusion of the right thigh by the fall of some heavy substance from the explosion which occurred at one A.M., after the Russians left the Redan.

There is not space to follow the details of the treatment of this case. The cure was protracted by large and troublesome bed-sores; and attention to these, to the discharges from the wound, and preserving favorable position, occupied much time and care daily, and caused many changes in the appliances for these objects to be from time to time necessary. On November the 4th, union had so far taken place that he was able to raise his body from the knee upward while in bed, without apparent motion at the seat of fracture. On November 15th, in consequence of the great explosion at the right siege-train, he had to be carried to another division of the camp; this was effected without harm. In the middle of January he was able to sit in a chair without inconvenience; and on February 22d he left the Crimea for England, being able to walk with the assistance of crutches. Union was then firm; but a slight serous oozing continued from the wound of exit, and there was much stiffness of the ankle and knee joints from the long-continued constrained position to which he had been subjected. In July, 1856, after his arrival in Ireland, indications of pus collecting manifested themselves at the wound of exit; and Professor Tufnell, on passing a bougie about seven inches in the course of the wound, evacuated a small abscess, and felt a piece of bone trying to make its way to the surface. This was subsequently removed, and, under

Mr. Tufnell's able care, the stiffness of the joints gradually disappeared, and he was enabled to return to duty.

Dr. Macleod says that, after many inquiries respecting cases of this nature in the hospitals of the other armies engaged in the war, excepting one presented by Baron Larrey to the Société de Chirurgie in 1857, he never could hear of any other but that of a Russian whose greatly shattered and deformed limb he often examined.* It had united almost without treatment. Two cases of united fractures of the femur in the upper third have arrived from the late mutiny in India, and in both, Dr. Williamson records, a good and useful limb had resulted, one with shortening of $1\frac{1}{2}$, the other $3\frac{1}{2}$, inches. Still more recently, M. Jules Roux, of the St. Mandrier Hospital, at Toulon, has given a list of no less than twenty-one cases of gunshot injuries of the upper third of the femur, which he had examined on their return from the Italian war of 1859, in all of which consolidation of the fracture had taken place. We have no data by which we can estimate the proportion of these cases of union to those in which other results ensued.

The proportion of recoveries in amputations in the upper third of the femur in the Crimean war was under 13 per cent. Amputation at the hip-joint, both in the French and English armies, in all instances proved fatal. The two patients who survived the longest were operated on by the late Director-General after the battle of the Alma: one, a soldier of the 33d Regiment, died at Scutari three weeks after the operation; the second, a Russian, died on the thirtieth day after, from "extensive sloughing and great debility."† One case of excision of the head, neck, and tro-

---

* Notes on the Surgery of the Crimean War, p. 264.

† In the surgical history of this war, this statement, which was quoted by the late Mr. Guthrie, in the Addenda to his Commentaries, is said to be a mistake, on account of the absence (not to be wondered at, amid the confusion of that period) of official records on

chanter of the femur in the Crimea recovered, operated upon by Dr. O'Leary; the only known successful case of excision of the hip-joint after a gunshot wound. The operation was performed on the same day that the wound was received. In the Sleswick-Holstein campaigns, amputation at the hip-joint was performed seven times; one patient only survived, a young man, aged seventeen years, operated upon by Dr. Langenbeck. Resection of the upper part of the femur, including the head and two inches below the small trochanter, was performed once, but the patient died from pyemia. At the post-mortem examination, the right shoulder and ankle joints were found to be filled with pus. The operation in this instance was performed three weeks after the injury. No case of amputation, nor of resection, at the hip-joint has returned from the Indian mutiny. M. Legouest, in a recent essay in the *Memoirs of the Society of Surgery*, at Paris, maintains that amputation at the hip-joint should be reserved for cases of fracture with injury to the great vessels, and that where the vessels have escaped, resection should invariably be performed. He also inculcates, as a general principle, not to perform immediate *primary* amputation at the hip-joint in any case; but, even in the severest forms of injury, to postpone the operation as long as possible.* For the *consecutive* results of gunshot

---

the subject. Special reports on these cases were obtained at the time from Scutari, and were shown to the writer by the late Director-General shortly before his decease.

* A committee was appointed by the Surgical Society of Paris to examine and report upon this essay of Dr. Legouest on Coxo-femoral Disarticulation for Gunshot Wounds. Baron Larrey drew up the report, which will be found in the 5th vol. of the Mémoires de la Société de Chirurgie, 1860. It confirms the principle laid down by Dr. Legouest, excepting only those cases of fracture where the mutilation of the limb from a heavy projectile has been so great as to partly separate it from the pelvis, and those in which there has been simultaneous lesion of the crural vessels and femur near the pelvis, with extensive laceration of the surrounding tissues.

wounds, the operation presents a less unfavorable aspect
than for immediate injuries. M. Jules Roux has recently,
at Toulon, performed amputation at the hip-joint six times
for the consequences of wounds received during the war in
Italy, and of these, four have been successful.

With regard to gunshot fractures in the middle and lower
third of the femur, the experience of the French and English
armies in the Crimea has tended to confirm the doctrine of
the older military surgeons, that many lives are lost which
might be otherwise preserved, by trying to save limbs; and
that, of the limbs preserved, many are little better than in-
cumbrances to their possessors. In the late Italian battles,
the practice of trying to save lower extremities, after comminuted fractures in these situations of the thigh, appears to
have been abandoned. Eight cases of union after compound
gunshot fractures of the femur in these situations have, however, returned from the late mutiny in India; and this is a
much larger proportion than was that of the recoveries from
the Crimea. Dr. Williamson, who records these cases, is
inclined to attribute this success in a great measure to the
use of dooleys for the conveyance of wounded, and argues
that it would be advantageous to introduce them into Euro-
pean warfare. But wounds generally, where proper care is
taken, heal more favorably in southern latitudes, east or
west, probably owing to the climate admitting of so much
more free an access of fresh air by day and night to the
patient than can be afforded, without inconvenience, in
colder or more variable climates. The dooley is most advantageous and comfortable as used in the East, where it is
an ordinary mode of conveyance among all classes, and the
bearers — a special race in each Presidency — are trained
from childhood to the occupation; but, from experience of
the peculiar habits and tenets of these men, both Madrassees
and those of Bengal, it seems scarcely probable that they
would prove efficient, even if they could exist, or that their
wants could be provided for in the numbers necessary to be

serviceable, with armies in northern latitudes. French surgeons have remarked how much more favorably, *cæteris
paribus*, wounds heal in Algeria, where they have only the
same kinds of conveyance for wounded as in Europe; and
the difference is accounted for by the favorable influence in
this respect of a warmer climate.

In fractures of the leg, where neither the knee nor ankle
joints are implicated, the results of conservative attempts
have been more favorable. In the Crimea, the recoveries
without amputation being resorted to were: in fractures of
both bones, nearly 19; tibia only, 36·3; fibula only, 40·9
per cent. When the fracture is comminuted, and implicates
the knee or ankle joint, opening the capsule, amputation is
necessary. The knee-joint was once excised in the Crimea,
but the patient died; as was the case in the only other instance where this operation is known to have been performed
for gunshot injury in the Sleswick-Holstein campaign. In
the treatment of fractures of the leg, where it has been determined to seek union, the same remarks apply as those
made above in respect to fractures in the upper extremity.
In wounds of the foot it is especially necessary to remove
as early as possible all the comminuted fragments of the
bones injured, or tedious abscesses and much pain and
constitutional irritation are likely to ensue.

AMPUTATION.

It is not necessary to refer at much length to the question
which was formerly disputed upon—the advantages of *primary* as compared with *secondary* amputation in gunshot
wounds—for military surgeons, whether acting at sea or on
land, have practically determined the subject. For a long
time the directions of John Hunter, that amputation should
not be performed until the first inflammation was over, based
on the argument that the "amputation is a violence super-

11

added to the injury, and therefore heightens the danger,"
and that this danger is aggravated in the instance of a man
laboring under mental agitation, as on the field of battle,
had great weight among English surgeons; but experience
has led to a different practice.   The greater success of primary amputation appears to be attributable to the facts,
that a contused and mangled limb is a constant source of
accumulating irritation; that the exciting circumstances
connected with battle lead a man to bear with courage at
an early stage what subsequent suffering and anxiety may
render him less willing to submit to; that a soldier, when
first wounded, is most probably in stronger health than he
will be after hospital restraint and confinement; that though
the amputation is a violence, it is one the patient is likely to
submit to with resignation, knowing that it is performed to
remove parts which, if unremoved, will destroy life; and
lastly, because the operation takes away a source of dread
which must weigh down the sufferer so long as it is impend-
ing.   The present practice has resulted from testing both
modes of amputation.   Mr. Guthrie showed, from the experience of the Peninsular war, that the loss in secondary
amputations had constantly exceeded that from primary
amputations in both the upper and lower extremities.   More
recent observations in both English and French campaigns
have confirmed this result.   Dr. Scrive records that the experience of the French army in the Crimea showed the success
of primary amputation sometimes exceeded by two-thirds
that of secondary amputation.   He excepts amputations at
the hip-joint, and cites, as his reason for this exception, that
in nine cases where the hip-joint amputation was performed
primarily, death followed the operation a few instants or a
few hours afterward; while in three cases which he witnessed, where the amputation was consecutive, one lived
five, another twelve, and the third twenty days.   In respect
to the particular time at which primary amputation is to be
performed, the general practice of the present day is, when

the operation is inevitable, to perform it as soon as it can be done; provided the more intense effects of "shock," where it has supervened on the injury, have passed off; and this practice generally accords with the feelings of soldiers, who not unfrequently press the surgeon for an early turn in being relieved from the suffering resulting from a shattered limb. In the cases where primary amputation is to be performed, a further reason given by Dr. Scrive for the operation being done on the same day that the wound is received is, that chloroform acts then so much more benignantly and readily; while, on the following day, or day after, traumatic excitement becomes very energetic, and considerable resistance is offered to its influence by wounded men, and longer time and a much larger dose of the chloroform are required to produce the state of anesthesia. If only a moderate amount of "shock" exist, this does not appear to be a sufficient reason for delaying amputation; for a moderate exhibition of stimulus and a few consolatory words will often remove this, and, even though some faintness, pallor, and depression remain, no ill consequences ensue. The late Director-General, in a letter to the late Mr. Guthrie, written in 1855, mentioned the case of a soldier of the 90th Regiment, whose right arm he removed at the shoulder-joint on the 10th of July, for great destruction of soft parts and extensive injury to the bone: "The patient was so low when placed on the table that brandy and water were given to him, and he was then immediately afterward placed under chloroform. When I had finished, it was observed that his pulse was stronger than before the operation." This man recovered without a bad symptom, and is now one of the Commissionaires in London. Indeed, in the Crimea, primary amputations were repeatedly performed where shock had not wholly disappeared, and no harm resulted from the practice. The introduction of chloroform, by its negative operation of preventing pain or alarm, and by its positive action as a stimulus, has done much to remove many of the

objections which were urged by John Hunter against early
amputations after gunshot wounds.   If collapse be intense,
more than is accounted for by the wound to the extremity,
suspicion will be excited that some internal injury has been
also inflicted, and delay will be necessary for further observation of the patient.   When active operations are proceeding,
and it is necessary to carry the wounded to any distance, the
advantages of early removal of shattered limbs are obvious.

SECONDARY HEMORRHAGE.

Army surgeons meet in practice with secondary more frequently than primary hemorrhage in gunshot wounds.   It
may arise in several ways.   Sometimes it results from the
coagulum being forced out of an artery in which hemorrhage
had previously been spontaneously averted by the ordinary
natural process, this accident being consequent upon muscular exertion or increased impulse of the circulating system
from any cause.   This occurrence in the bottom of a deep
wound will be often found to be a very troublesome complication.   Sometimes an artery which did not appear to be
injured in the first instance ulcerates or sloughs; or, without
direct injury, a vessel may become involved in unhealthy de-
terioration of the wound, and give way; or, in a granulating
wound, general capillary hemorrhage may be excited by
stimulus of any kind, such as venereal excitement or excess
in drinking; or the coats of the vessel may ulcerate under
pressure from a detached fragment of bone or from some
foreign body; or the artery may be accidentally penetrated
by the end of a sharp spiculum.   Secondary hemorrhage
has been said to arise from increased arterial action, from
the first to the fifth day; from sloughing, the effects of contusion, from the fifth to the tenth; from ulceration, to any
more distant date.   M. Baudens has remarked that he has
observed secondary hemorrhage to be most frequent about
the sixth day after the wound—the traumatic fever having

then reached its highest point of intensity, and the sharp, hurried contractions of the heart having most power in forcing out the coagula. If we could compare all the cases of hemorrhage which occur, secondary would, perhaps, statistically appear less dangerous than primary hemorrhage; for the latter, when happening from large vessels, must be very generally fatal, while, when hemorrhage occurs in them secondarily, the collateral branches have become partially adapted to the interruption of the flow of blood through the regular channel. Moreover, the larger arteries, when once filled with coagula and well contracted, fortunately do not frequently yield to the impulse which serves to produce secondary hemorrhage in vessels of smaller caliber.

Secondary hemorrhage is not uncommon after deeply-penetrating gunshot wounds of the face, and sometimes it is difficult to determine the site of the bleeding vessel. It may be so situated that the rule of tying both ends of the bleeding artery in the wound cannot be carried out, and where, if the ordinary styptics fail, resort must be had to the ligature of the common trunk from which the bleeding vessel branches. In the museum at Fort Pitt is a cranium showing the passage of a musket-ball from the inner side of the right orbit to the entrance of the carotid canal in the petrous portion of the temporal bone, where the ball had lodged. Death ensued, ten days after the wound, by hemorrhage from the internal carotid. In another case, a branch of the external carotid artery was wounded by a ball which penetrated at the zygomatic fossa. Secondary hemorrhage ensued, and the usual means failed to arrest it. The external carotid was tied; but blood continued to flow, though less abundantly than before. Compression in the wound, which failed previously, now served to arrest the hemorrhage, and cure followed. Care must be taken, before tying the trunk, that pressure upon it exerts control over the hemorrhage from the wound; for the irregular course of projectiles is not unlikely to lead to mistakes, such as tying the common ca-

11*

rotid, which is stated to have been done when the hemorrhage has been from the vertebral artery.

The rule of treatment, however, holds good in secondary as in primary hemorrhage—the bleeding vessel must be secured at the wounded part whenever practicable, and it must be tied both above and below the line of division, taking care to ascertain that the spot where each ligature is applied is sound. Hemorrhage from general oozing, from sloughing, and other causes must be treated on the general principles applicable in all such cases.

WOUNDS OF NERVES.

Temporary paralysis from contusion of a nerve in the passage of a projectile is not unfrequent. Complete loss of power of motion and sensibility in a limb occasionally follows gunshot injuries, and generally indicates complete division of the nerve. Instead of complete paralysis, there may remain only modified deprivation of sensibility, partial loss of muscular force, and diminished power of resisting cold, with or without pain; and these symptoms may either be the result of contusion, with the effects perhaps of inflammatory action or of partial division. When a foreign body is lodged in or among nerves, it may induce tetanic symptoms of a fatal character, or great irritation and intense pain may result; and unless the source of these latter symptoms can be found and removed, if in a large nervous trunk of one of the extremities, they will sometimes lead to the necessity of amputation. The gunshot injuries which cause division of large nerves, however, are usually attended with so much destruction of other parts that the question of amputation has scarcely ever to be considered in reference to lesions of nerves alone. Atrophy of tissues and contractions of muscles are common results of injuries to nerves from gunshot, and often lead to soldiers being disabled for further service. Occasionally, after severe injuries, the functions of sensation and power of motion gradually return, in some instances with perfect cure, but mostly with impaired power of resisting rapid alternations of temperature, especially cold. A case is mentioned in the Surgical History of the Crimean War where a soldier had the right sciatic nerve severely injured by the passage of a musket-ball. Total loss of sensation in the right foot followed. The wound was healed a month after it was received, and sensation slowly returned in the foot; but the restoration was attended with intensely burning pain, unrelieved by any applications. Gradual recovery took place. Dr. Williamson's returns show eight cases of gunshot wounds with direct injury to nerves among the men invalided from India, after the late mutiny; all were wounds involving the brachial plexus, and in all there was paralysis, partial or complete, of the upper extremity on the injured side. In one case, the loss of function appears to have been almost confined to the hand; all the fingers were fixed in a straight position, and numb, and any attempt at bending them occasioned intense pain in the course of the median nerve. The hand was cold and affected with nervous tremor, but the motor power and sensibility of the thumb were preserved. The following hitherto unrecorded case illustrates several points: A soldier of the 37th Regiment was wounded at Azinghur, on the 27th of March, 1858, by a musket-ball, through the right side of the neck. It entered just below the horizontal ramus of the jaw, and made its exit behind, over the scapula. About three pints of blood escaped, supposed to be from the external jugular vein. The wound healed favorably, but he lost the use of his right arm, at first completely, and afterward partially, for three months. At the expiration of that period the power of the arm was restored; but he was invalided home on account of severe pain in the back of the neck, "resembling toothache," which all treatment failed to relieve. This pain spontaneously and gradually ceased; there is still some loss of substance of the trapezius muscles of the right side of the neck, and of the right as compared with the other arm, with occasional numbness when the man is in heavy marching order; but in all other respects he is well, and is at his regular duty.

TETANUS.

One cause of fatal termination in gunshot wounds is tetanus. It is generally believed that the proportion of deaths from this source is greater after actions in tropical climates, and that exposure to the night air in such regions has some especial effect in producing them. The most common cause appears to be, however, the local injury to nerves, already mentioned, producing irritation along their course, and so leading to some morbid condition of the ganglionic portions of the motor tracts of the spinal cord. In the Crimean campaign, the proportion of tetanus was remarkably small as compared with former wars, being, according to the returns, only 0·2 per cent. of the number wounded. Dr. Scrive records that not more than thirty cases of tetanus occurred among the French wounded during the whole Crimean war, and this would show a somewhat less ratio even than in the British army. Dr. Stromeyer records only six cases of tetanus among 2000 wounded in the campaign of 1849 against the Danes. Three of these, in which the disease assumed a chronic form, recovered. There was only in one case injury of bone. Warm baths and opium were the remedies in the successful cases.

Sir G. Ballingall made the calculation that one in seventy-nine is the average number of tetanic cases among wounded, and states that the proportion of recoveries is so small as scarcely to be taken into account. Three cases occurred to the writer, in the Crimea, after gunshot wounds; all proved fatal. In one there was a severe fracture of the ischium and injury of testicle by grape-shot. In a second, a rifle-ball entered just above the left knee, and lodged. Eight days after the injury an abscess was opened near the tuberosity of the ischium, and the ball was removed from that spot. The same day tetanus set in, and he died three days afterward. The ball had injured the sciatic nerve, which was found to be reddened superficially; while the neurilema, also, under an ordinary magnifying-glass, showed indications of inflammation. A piece of cloth was found lying midway in the long, sinus-like wound made by the ball. In a third, the bullet passed through the axillary region. The patient progressed favorably for some days, when tetanic symptoms appeared, and under these he sank. At the post-mortem examination, some detached pieces of woolen cloth were found lying entangled among the axillary plexus of nerves. Twenty-one cases altogether supervening on gunshot injuries are shown in a table in the Crimean records. Of these, ascertained injuries to nerves by projectiles, or division of nerves by amputation, occurred in eleven cases; three followed compound fractures, and seven flesh wounds. The average period at which the tetanic symptoms appeared was eight days and a half after the receipt of the injury; their duration prior to death, three days and a half. One case only recovered—a soldier of the 93d Regiment, wounded in the right buttock by a shell explosion. A fragment nearly a pound in weight was removed soon after the injury. Seventeen days after trismus set in, when a further examination of the wound led to the discovery of an angular fragment of shell which had been previously overlooked. It was deeply lodged, and resting on the sciatic nerve. On removing this, which weighed eighteen ounces, the sheath of the nerve was seen to be lacerated to nearly one inch in extent. Calomel and opium were now given; salivation appeared three days afterward, the trismus subsided, and the man gradually convalesced.

Beyond the extraction of any foreign bodies which may have lodged, as in this last case, it is not known that there are any indications for special treatment of tetanus as occurring after gunshot injuries. The employment of woorali has

again been brought into notice by its successful administration by M. Vella, of Turin, in the case of a French sergeant
wounded in the metatarsus of the right foot, on the 4th of
June, 1859, at the battle of Magenta, by a musket-ball which
lodged. The projectile was extracted three days after his
admission into hospital at Turin, on the 10th of June, and
tetanus set in three days afterward. But the woorali failed
in two other cases; and it has yet to be determined, should
it be found to possess any peculiar power over tetanic spasm,
to what class of cases its properties are applicable.

Hospital gangrene, a common disease of wounded soldiers when circumstances of war lead to overcrowding in ill-
ventilated buildings, and to deficiency in the proper number
of attendants for securing personal cleanliness and purity of
atmosphere, with inferior diet ; and **Pyemia,** a frequent cause
of fatal termination after gunshot fractures, injuries of joints,
and other suppurating wounds, especially under the influence
of circumstances like those above named, are treated separately under their respective heads.

ANESTHESIA IN GUNSHOT WOUNDS.

The complete applicability of chloroform on the field to
injuries caused by gunshot, as to all others in civil practice,
is established among Continental surgeons, and among a
majority of British army surgeons. The first opportunity
of testing chloroform largely as an anesthetic agent in British military surgery occurred in the Crimean war, and a long
report on the subject will be found in the published Surgical
History of the Campaign. The general tenor of this report
is to limit considerably the use of chloroform—in minor operations, on the ground of occasional bad results, even when
the drug is of good quality and properly administered; or,
in cases where the shock is very severe, on the ground that
such do not rally, owing to the depressing effect of the drug,
after the anesthesia has gone off; or in secondary opera-

tions, from the systems of the patients having been much reduced by purulent discharges. But from the report it appears that only one patient died from the effects of chloroform ; and in this instance, Professor Maclagan, of Edinburgh, to whom a portion was forwarded for examination, reported the drug to be "acrid and nauseous when inhaled," and "totally unfit for use." On the other hand, Dr. Scrive, chief of the French Medical Department in the East, has written, in his Relation Médico-Chirurgicale de la Campagne d'Orient, p. 465 : "De tous les moyens thérapeutiques employés par l'art chirurgicale, aucun n'a été aussi efficace et n'a réussi avec un succès aussi complet que le chloroforme ; jamais, dans aucune circonstance, son maniement sur des milliers de blessés n'a causé le moindre accident sérieux ;" and, more recently, Surgeon-Major M. Armand has written : "During the Italian war, chloroform was as extensively used and was as harmless as in the Crimea. I never heard of an accident from its use."

At the commencement of the Crimean war, the Inspector-General at the head of the British Medical Department circulated a memorandum "cautioning medical officers against the use of chloroform in the severe shock of serious gunshot wounds, as he thinks few will survive where it is used ;" but as far as chloroform was available, it was used by many medical officers from the commencement of the campaign, and its employment became more general as the campaign advanced. It was constantly used in the division to which the writer belonged throughout the war ; and no harm was ever met with from its use, while certain advantages appeared especially to fit it for military surgical practice. So far from adding to the shock of such cases as an army surgeon would select for operation, the use of chloroform seemed to support the patient during the ordeal ; and the writer has several times seen soldiers, within a brief period after amputation for extensive gunshot wounds, and restoration to consciousness, calmly subside into natural and refreshing sleep.

One reason for not using chloroform in the Inspector-General's caution was, that the smart of the knife is a powerful stimulant ; but "pain," it has been remarked by a great surgeon, "when amounting to a certain degree of intensity and duration, is itself destructive ;" and there can be little doubt that the acute pain of surgical operations, superadded to the pain which has been endured in consequence of severe gunshot fractures, has often, where chloroform has not been used, intensified the shock, and led to fatal results. In civil surgery, statistical evidence has demonstrated that the mortality after surgical operations has lessened since the use of chloroform; and it is believed the same result would be shown, if opportunity existed, in army practice. In the report of a case in the Crimea, instancing, perhaps, the greatest complication of injuries from gunshot of any which recovered, Dr. Macleod remarks casually in his Notes, p. 265 : "This amputation was of course done under chloroform, otherwise it is questionable whether the operation could have been performed at all, the patient was so much depressed." Mr. Guthrie, in the Addenda to his Commentaries, remarked, from the reports and cases which had reached him, that chloroform had been administered in all the divisions of the army save the second, and had been generally approved; and that the evidence was sufficient to authorize surgeons to administer it even in such wounds as those requiring amputation at the hip-joint. The late Director-General amputated in three instances at the hip-joint, after the battle of the Alma, under chloroform—two on the 21st and one on the 22d September—and all these lived to be carried on board ship on the latter-named day, and two, as before stated, lived several weeks. The absence of increased shock from pain during the amputation very probably enabled these patients to withstand the fatigue of removal to the coast and embarkation on board ship. With regard to the objection of occasional bad results, a recent estimate has shown that the probable proportion of all the deaths which have occurred from chloroform to the operations performed under its influence, exclusive of its use in midwifery, dental surgery, and private practice, has been one in 16,000; and as these accidents may equally occur during "minor operations," in army practice as in civil life, it should be used or not at the option of the patient.

In respect to the danger of anesthetics in the secondary operations connected with gunshot wounds, Dr. Scrive's experience has led him to remark: "When consecutive amputation is rendered necessary by the gradually increasing debility of a wounded man from purulent discharges, chloroformization takes place with the most perfect calm on the part of the patient;" and he classes its use under "chloroformization de nécessité." The general rules followed in civil surgery must be equally applicable in these cases.

It must frequently happen in military practice that several operations have to be performed in rapid succession on the same person, from necessity of a speedy removal of the wounded; and, moreover, from the number of cases which are suddenly thrown on the care of the army surgeons after a general engagement, it must frequently occur that the diagnosis of a case is more or less doubtful. In such instances, the use of chloroform, by diminishing pain and preventing shock, and thus giving the opportunity of more accurate examination of parts, becomes particularly valuable in army practice. After the battles of Alma and Inkerman, when orders were given to remove the wounded as speedily as possible, the first-named consideration frequently occurred. The case of Sir T. Trowbridge is quoted by Mr. Guthrie. This officer had both feet completely destroyed by round shot at Inkerman, and it was necessary to amputate, on one side at the ankle-joint, on the other in the leg: the use of chloroform enabled the two operations to be performed within a few minutes of each other with perfect success. The amputations were done by the late Director-General of the Army Medical Department. In illustration of the second

casualty, the following, which happened to the writer at
Alma, may be named. A man of the Grenadier company
of the 19th Regiment had a leg smashed by round shot. It
was a question whether the fracture of bone extended into
the knee-joint. Two superior staff-surgeons were near; a
hasty consultation was held, and it was decided that the
probabilities were in favor of the joint being intact. Amputation was performed, and the tibia sawn off close to the
tubercle. It was then rendered evident that there was
fissured fracture into the joint. As soon as the man had
recovered from the state of anesthesia, the necessity of amputation above the knee was explained to him, and he readily
assented. This was shortly afterward done, and the man
recovered without any unusual symptoms, and was invalided
to England. It is not likely, without chloroform, in a doubtful case of this kind, that the chance of saving the knee would
have been conceded.

In the British army in the Crimea chloroform was generally applied by simply pouring a little on lint. The chief
objection against this in the open air is probably the waste
which is likely to be occasioned. Dr. Scrive says it always
appeared to him most advantageous to use a special apparatus, as well to measure exactly the doses, as to guarantee
a proper amount of mixture of air; and that although he
never saw a fatal result, he had several times seen excess of
chloroformization from the use of lint rolled up in the shape
of a funnel. The instructions which he gave were, never to
pass the stage of strict insensibility to pain, never to wait
for complete muscular relaxation; and to this direction being
carried out he attributes the fact that no death occurred
from chloroform in the French army in the Crimea. In an
article on anesthetics, in the *Medico-Chirurgical Review*,
October, 1859, Dr. Hayward, of Boston, has strongly advo-
cated the use of sulphuric ether above all other anesthetics.
The quantity required to produce anesthesia—from four to
eight ounces—would render the use of this agent almost
impracticable in extensive army operations in the field.

AFTER-USEFULNESS OF WOUNDED SOLDIERS.

The results of wounds unfit soldiers for military service in many ways, according to the nature of the wound and the region in which it is inflicted; and the pensions consequent on their discharge entail heavy expenses of long duration on the country. It was hoped that the improvements in conservative surgery would have diminished the number of disabled soldiers as compared with former wars; but the corresponding improvements in the power and means of destruction, with other circumstances, have defeated this hope, and the returns do not show such to be the result. Even the cases where resections of the joints have been performed, and fractures united, which previously would have been treated by amputation, have rarely presented such cures as to render the men available for military service, though the preserved limb may still be of use in the work of civil life. Formerly, all men who thus became unfitted to perform any of the duties to which a soldier is liable were removed from the army; but, by an order from the Horse Guards of 1858, wounded soldiers, though rendered unfit for active service in the field, were directed to be retained for modified duty in such employments as they are capable of executing. The results of the increased practice of conservative surgery may, therefore, prove valuable to the public service, now that the opportunity of secondary employment is laid open. The reports from the hospitals in Italy show that during the recent campaign in that country the practice of conservative surgery after gunshot fractures has been very limited, and in the lower extremity has been almost wholly abandoned, early amputation being practiced instead.

It is believed, that should England become again involved in war, a greater amount of systematic scientific observation will be brought to bear upon the subject of gunshot wounds than circumstances have ever previously admitted. Hitherto, the majority of the younger medical officers of the army have found themselves, on the occasion of war, suddenly in possession of a large number of wounded officers and soldiers to treat, with only those general principles of surgery to guide them which they had originally obtained in their studies in civil hospitals and schools; but this knowledge, essential and absolutely necessary above all other as it is, has been long admitted in the first-class powers of the Continent, whose military experience is necessarily greatest, to be incomplete for this purpose. Now that an Army Medical School has been established in England, and that in it the large number of sick and wounded who annually return from all parts of the world—serving to illustrate, among other subjects, the consequences of wounds and of the surgical operations performed for them in all their varieties—will be turned to account, as well as the great collection of preparations in the museum of the Army Medical Department, it is only reasonable to hope that the opportunities of study in these specialties which will be afforded to every medical officer at his entrance into the army will cause each individual, not only to be ready to apply at any moment all the improvements derived from experience and observation, up to the most advanced period, in this branch of the profession of surgery, but will also best prepare the members of the department for extending still further the sphere of usefulness which has been cultivated by their predecessors.

THE END.

www.ingramcontent.com/pod-product-compliance
Lightning Source LLC
Chambersburg PA
CBHW021934190326
41519CB00009B/1023